Ameer Hamza Khan · Xinwei Cao · Shuai Li

Management and Intelligent Decision-Making in Complex Systems: An Optimization-Driven Approach

Springer

Ameer Hamza Khan 🆔
Department of Computing
Hong Kong Polytechnic University
Hong Kong, Hong Kong

Xinwei Cao
School of Management
Shanghai University
Shanghai, China

Shuai Li
School of Information Science
and Engineering
Lanzhou University
Swansea, UK

ISBN 978-981-15-9391-8   ISBN 978-981-15-9392-5 (eBook)
https://doi.org/10.1007/978-981-15-9392-5

This Springer imprint is published by the registered company Springer Nature Singapore Pte Ltd.
The registered company address is: 152 Beach Road, #21-01/04 Gateway East, Singapore 189721, Singapore

Preface

Management is about the manipulation of information, including planning, organizing, directing, and controlling, for a process under given input to reach a certain amount of optimality in output. Management is everywhere, from marketing to politics, from individuals to society, from natural science to social science. Better management grants an opportunity for managers to make better decisions. This is especially important to complex systems, as existing theories are mostly developed to benchmark problems traditionally described by concise and simplified mathematic models. In the era of big data, The huge amount of data and the advance of big data analytic techniques give us new insights into the resolution of complex system management and decision-making. This book provides a timely report on our results along this research direction, with a goal to embrace complexity with big data.

In recent years, articulated agents have been widely studied and have found increased research attention from academia as well as from industry. An articulated agent consists of several components; base-platform, joints, links, and end-effector. Studying the design and motion of each of these components has been extensively performed. In particular, the formulation of advanced management techniques, to accurately, robustly, and safely manage the motion of these components is of great interest to researchers. For example, developing high-level management algorithms which can help the articulated agent to navigate through the environment without collision with the surrounding safely is one of the examples of goals researchers try to achieve through advanced sensing mechanism and algorithms to manage these systems. Additionally, the technology related to the design and fabrication of end-effectors has also made significant progress. For example, the end-effector made from flexible and soft material has been gaining spotlight because of its inherent safety advantages. The same level of safety is hard to achieve through traditional rigid end-effectors and requires sophisticated sensing and management algorithms. In addition to that, the recent paradigm development is control theory, such as impedance control. It has also found its application in every robotic system in general and articulated agents in particular.

In this book, we focus on three aspects related to the development of articulated agents: presenting high-level control algorithms for obstacle avoidance of articulated agents in industrial environments, experimental study of the properties of soft robotic agents as the end-effectors of a rigid articulated agent, and low-level torque-control loop to accurately control the joints of the articulated agents. This book is divided into the following four chapters.

Chapter 1—This chapter presents a novel motion planning and decision-making strategy for managing the functions of a redundant articulated robot, called Beetle Antennae Olfactory Recurrent Neural Network (BAORNN). The proposed approach simultaneously manages the motion of the robot according to the desired path while avoiding any obstacle present in the surrounding of the robot. The proposed strategy is based on the optimal-control approach and unifies the task of obstacle avoidance and trajectory tracking into a single optimization problem using the penalty-term method. Another feature of the proposed algorithm is the straightforward formulation of the penalty term using the following principle: maximize the minimum distance between the articulated robot and obstacle. We used the GJK (Gilbert–Johnson–Keerthi) algorithm to calculate the gap between the articulated robot and the obstacle by using the three-dimensional geometry without making any assumption about their shapes and sizes (i.e., assuming them to be point objects), implying that our algorithm generally deals with arbitrarily shaped robots and obstacles. Simulation results using LBR IIWA, 7-DOF articulated robot, are presented to analyze the performance of the proposed framework.

Chapter 2—This chapter presents an experimental study of structurally uncertain soft agents, also known as soft robots. Due to structural softness and flexibility, the mathematical model of these soft agents is highly nonlinear, and an infinite degree of freedom (DOF) is required to model their behavior exactly. Currently, model-free controllers, such as Proportional Integral Derivative (PID), are the most commonly used control strategies for controlling the soft agents. This chapter presents a systematic experimental study to comprehensively characterize the behavior of the PID controller for the soft agents and identify their unique properties as compared to rigid-robots. In this chapter, we analyzed the behavior of three variants of the PID controller. We studied their performance for the case of manual tuning, using the Ziegler–Nichols method, as well as automatic tuning, using coordinate descent. The experimental results statistically demonstrate the efficacy of the proposed automatic tuning algorithm. Additionally, we empirically showed that, for the case of soft agents, the PID controller essentially reduces to the PI controller.

Chapter 3—In this chapter, we discuss another issue being faced for the control of pneumatically actuated soft agents. Soft agents undergo vigorous oscillation, when deactivated, before settling down because of their flexible bodies. These oscillations might compromise the structural integrity of a soft agent with time. In this work, we present a novel design of a 6-chambered parallel soft agent and propose an effective active damping method by a smart distribution of the 6 actuation chambers. Experimental verification of the effectiveness of the proposed damping method is conducted on the proposed parallel soft agent. It is shown that the proposed method provides a high degree of oscillation damping, thus prolonging the actuator's life. Since the proposed method uses the components of the soft agent itself to create oscillation damping actively, there is no additional mechanical overhead.

Chapter 4—Motors constitute an essential component of articulated agents since they primarily control the motion of each joint. DC motors are one of the most common types of motors used in articulated agents. For the proper operation of an articulated agent, the lower-level control loops, i.e., the speed-control, current-control, and torque-control loops, must be well-formulated and able to track the reference signal accurately. In this chapter, we present an experimental platform consisting of two motors, mechanically coupled through the shaft, to study the simultaneous control of current and speed in DC motors. We propose the mathematical formulation of the kinematics and dynamics of the system and formulate a Proportional Integral (PI) controller combined with feedforward control law to control the current in the DC motor accurately. The experimental results presented in the chapter show that the bandwidth of the controller depends on the controller parameter and the filtering of the sensor value. If the filtering action is applied to the sensor value, the accuracy is increased; however, it decreases the bandwidth and increases the rise time of the controller. However, by appropriately selecting the filter, a compromise between bandwidth and accuracy can be achieved.

At the end of this preface, it is worth pointing out that, in this book, we have summarized recent advances for the articulated agents related to high-level motion planning and decision-making algorithms, end-effector technology, and optimization of low-level control loops. The motive behind the book is to trigger theoretical and practical research studies related to articulated agents. There is no doubt that this book can be extended. Any comments or suggestions are welcome, and the authors can be contacted via e-mail: shuaili@ieee.org (Shuai Li).

Hong Kong, Hong Kong   Ameer Hamza Khan
Shanghai, China   Xinwei Cao
Swansea, UK   Shuai Li
August 2020

Contents

Acronyms

BAORNN	Beetle Antennae Olfactory Recurrent Neural Network
BAS	Beetle Antennae Search
GJK	Gilbert–Johnson–Keerthi
PID	Proportional Integral Derivative
RNN	Recurrent Neural Network

Chapter 1
Obstacle Avoidance Based Decision Making and Management of Articulated Agents

Abstract Articulated agents, mainly comprise of two components: the mechanical structure and an end-effector to manipulate the objects. This chapter, along with chapter four of this brief, focuses explicitly on the first component, i.e., the mechanical structure of the articulated agent. This chapter introduces a novel algorithm for the planning, smart decision-making, and management of a redundant articulated robots, called the Beetle Antennae Olfactory Recurrent Neural Network (BAORNN). The proposed algorithm carries out the task of articulated robot's trajectory tracking and obstacle avoidance, simultaneously. It is a crucial feature for an articulated robot to conduct operations in an industrial setting without colliding with any nearby object. The proposed algorithm is based on an optimization-driven methodology and unifies the task of avoiding obstacles and monitoring into a single optimization problem using a penalty-term approach. A primary advantage of the penalty-term is that the optimizer is actively awarded for overcoming the barrier and, therefore, increasing the rate of convergence. Another feature of the proposed algorithm is that it also provides a simple formulation of the penalty term using the following principle: increase the minimum distance between the robot and the obstacle. In order to measure the distance between the articulated robot and the obstacles, we used the GJK (Gilbert-Johnson-Keerthi) algorithm by using three-dimensional geometry of the robot and obstacle without making assumptions of its shapes and sizes (i.e., assuming that they are point objects). For evaluating the performance of the proposed Framework, the results of simulations using LBR IIWA, 7-DOF articulated robot, are given.

1.1 Introduction

The issue of tracking and obstacle avoidance for a redundant articulated robot is targeted at computing an appropriate management strategy to guide the end-effector along a specified reference path while avoiding obstacles present in the environment.

© The Author(s), under exclusive license to Springer Nature Singapore Pte Ltd. 2021   1
A. H. Khan et al., *Management and Intelligent Decision-Making in Complex Systems: An Optimization-Driven Approach*, https://doi.org/10.1007/978-981-15-9392-5_1

Articulated robots have attracted growing research interest from academia [1–4] and from industry [5–11]. Industries are interested in using articulated robots to automate common tasks such as transporting, assembly, labeling, and food transportation. Accurate trajectory tracking is a vital condition for the industrial robotics [12–18], along with obstacle avoidance. To meet these requirements, redundant articulated robots [19] are particularly desirable because the extra degree of freedoms (DOFs) offered by redundant joints helps to achieve secondary design goals, such as preventing obstacles [20–25] It is well established in the literature that the problem of trajectory tracking and avoidance of obstacles in are challenging in itself [26]. Especially, because a closed-form inverse-kinematic model for most redundant manipulator configuration does not even exist [27, 28]. The integration of these two problems into one task poses a complex technical challenge.

The research literature has rigorously studied various aspects of industrial articulated robots. In addition to tracking algorithms, particular emphasis has also been put on designing optimal task-space trajectories for the articulated robot as well as analyzing the repeatability of these algorithms has been of great interest [29, 30]. For example, one of the standard algorithms, named Jacobian-matrix-pseudo-inverse (JMPI), was shown to have low repeatability [31]. Jerzy et al. [32] suggested a standardized method for calculating the repeatability of an industrial robot's pose and addressed additional factors impacting the optimal control strategy, such as mechanical and thermal pressure. Similarly, multiple algorithms were proposed [33] to improve the repeatability of the articulated robot during the long-term service. Other methods for improving the articulated robot's repeatability include the learning algorithm to approximate the kinematic model in real-time [34]. The learning algorithm dynamically adapts to and compensates for variability in the machine structure in real-time. Likewise, Visual Servoing related methods have also been suggested [35] to use computer vision algorithms to enhance the control of industrial articulated robots.

Kinematic tracking of a redundant articulated robot is a well-studied topic in the [19, 22, 36] robotic literature. Imagine an industrial articulated robot, for example, allocated to push an item from one position to another by executing a given route within the cartesian task-space. There are infinite numbers of trajectories in the joint space for a redundant articulated system, to track a specified trajectory in cartesian space. Traditionally, Jacobian-matrix-pseudo-inverse (JMPI) [37] is used to resolve redundancy. Nevertheless, JMPI can only be used to address the equality constraints. Furthermore, it can not accommodate obstacle avoidance, which is usually represented as inequality restrictions [21, 26]. Also, as later shown by Klein et al. [31], the JMPI does not produce repeatable results and can potentially lead to undesirable joint configuration. Additionally, the estimation of Jacobian's pseudo-inverse is a computationally comprehensive task. Current approaches to redundancy resolution model the kinematic tracking as a constrained optimization problem [20–22]. Such optimization-centered methods are capable of solving additional inequality constraints along with the trajectory tracking [14, 38]. For instance, Wei et al. [39] and Wang et al. [40] used this to control flexible joints on articulated robots. Li et al. [36] introduced a dual Recurrent Neural Network (RNN) to solve the problem of monitoring optimization for several real-time articulated robots. Ding et al. [41] used this method to minimize the joint torques and constraint the joint-angles in a mechanically optimal range while accurately tracking the reference trajectory. Adaptive management techniques were also presented in the literature [42–49] which estimate the pseudo-inverse of Jacobian matrix in real-time instead of explicitly calculating its pseudo-inverse. The major advantage offered by optimization-centric approaches over adaptive control and JMPI algorithms is faster computation and ability to solve a general class of inequality constraints, which we leverage in this chapter.

Obstacle avoidance, along with trajectory tracking, is also an essential primary objective for an articulated robot [26, 50]. The industrial robots often have to work in a complex environment and communicate with other robots and structures in the surrounding environment. A traditional strategy, as proposed by Khatib [51], was the conception of the "artificial force field", which formulate goal position and obstacles as attractive and repulsive poles of a magnetic respectively. However, the concept of poles exists in the cartesian task-space, which requires inverse transformation into the joint-space before applying the necessary control signal. Similarly, Flacco et al. [52] used distance sensor to formulate an obstacles avoidance strategy. Guo and Zhang [21] presented an algorithm that takes joint-acceleration as input and tries to calculate a control signal which minimizes the joint-acceleration. Zhang et al. [26] proposed an obstacle avoidance strategy; however, they consider obstacles as point-objects which reduces the reliability and efficacy of their algorithm. A common feature of traditional methods is to introduce obstacle avoidance as a passive constraint in the optimization problem. Our proposed formulation overcomes this shortcoming by using a penalty term in the objective function, which actively rewards the optimizer for maximizing the distance between robot and obstacles. To summarize, the problems being addressed in this chapter are

1. Formulating a control algorithm for a redundant articulated robot to compute the necessary control actions in joint-space to follow a given task-space trajectory.
2. While following the given trajectory, the algorithm should satisfy the mechanical constraints, e.g., joint-angle limits.
3. The objects present in the surrounding of the articulated robot are considered obstacles, and collision with them should be avoided.

In this chapter, we leverage the fact that optimization-based approaches can incorporate arbitrary goal into the objective function [20, 21, 53–56]. We combine the problem obstacle avoidance and tracking control into a single objective function by using the penalty term. The proposed penalty term approach rewards the optimizer for avoiding the obstacle, in contrast to the traditional approaches, which simply incorporate it as inequality constraints [26]. This effectively reduces the problem to estimating the numerical solution of the optimization problem in real-time. Our formulated objective function has two goals: minimize the Euclidean-norm of tracking error and maximize the distance between the links of the robot and the obstacles. Instead of making an assumption about the shape of the obstacle or considering them as point objects, we used the Gilbert-Johnson-Keerthi (GJK) algorithm [57], to directly use the 3D-geometries of robot's links and the obstacle to estimate the required distance.

To efficiently estimate the numerical solution of the formulated optimization problem in real-time, we consider a metaheuristic approach; called Beetle Antennae Olfactory Recurrent Neural Network (BAORNN). We leverage the well known ability of metaheuristic optimization algorithms to solve complex optimization problems [58–62] in a numerically efficient manner. Our proposed algorithm is built upon on a nature-inspired metaheuristic algorithm, called Beetle Antennae Olfactory (BAO) algorithm [63–67], inspired by the food foraging behavior of beetles. BAO has shown practical for real-world applications as demonstrated by recently published results [68–72]. Specifically, the BAO algorithm allowed us to introduce the concept of the "virtual robots", which virtually anticipate the motion of a given control signal and only move the real robot when accuracy requirements are fulfilled. We present the formulation of the BAO algorithm as a Recurrent Neural Network (RNN), which enables fast prototyping and will be able to leverage the modern hardware developments related to neural computation.

The unique feature of the proposed algorithm is its formulation on position-level as opposed to velocity-level as presented by traditional approaches [26, 34, 36]. A key advantage of this approach is that it does not require the initial position of end-effector to lie on the reference trajectory. In contrast, the velocity-level algorithms explicitly require that the end-effector be manually moved to the initial point on the reference trajectory. Furthermore, the velocity-level algorithms are computationally expensive as they require the computation pseudo-inverse of the Jacobian matrix at each time-step. On the other hand, a position-level management algorithm avoids the mathematical manipulation of the Jacobian matrix, thereby significantly reducing the computation cost. It is also worth noting that the proposed algorithm does not assume the shape of the obstacle; neither consider it as a point object [26]. Instead, it directly uses the 3D-model of the robot to calculate the distance. As such, it works for any arbitrary shaped robots and obstacle, which makes it highly feasible for an actual industrial setup. The 3D-geometry of the obstacle can be easily estimated in real-time, given the modern depth mapping sensors and 3D construction algorithms. The main highlights of this chapter are:

1. We present a unifying framework based on an optimization-centric approach which combines the problem of tracking and obstacle avoidance for redundant articulated robots.
2. The proposed approach formulates the problem on position-level as compared to the velocity-level as done in most traditional works. The position-level algorithm avoids the manipulation and pseudo-inversion of the Jacobian matrix and resulting in a considerable reduction in the computation cost.
3. Using the GJK algorithm to estimate the Euclidean distance between the robot and an arbitrarily-shaped obstacle by directly using their 3D-geometries without making assumptions about their shapes.

4. We presented a recurrent neural network, called BAORNN, based on nature-inspired metaheuristic algorithm to solve the formulated optimization problem in real-time efficiently.
5. Extensive numerical analysis using a simulated model of KUKA LBR IIWA-14, a popular 7-DOF industrial robotic arm, are performed to demonstrate the performance of the proposed algorithm.

The remainder of this chapter is organized as follows: Sect. 1.2 presents the problem formulation. In Sect. 1.3, present BAORNN algorithm and its RNN architecture along with theoretical analysis. Section 1.4 outlines the simulation methodology, present the results, and discuss their implications. Section 1.5 concludes this chapter.

1.2 Formulation of Optimal Management

In this section, we will formulate the problem of tracking control and obstacle avoidance mathematically and unify it into one optimization problem.

1.2.1 Kinematic Tracking

Consider the task of moving a payload over a given trajectory using an articulated robot, say a circular path. Trajectory tracking requires the joint-space trajectory calculation, which will move the end-effector along the defined circular path. The position of the end-effector for a given articulated robot is a function of its joint angles. Consider, for example, an articulated robot m-DOF that works in a n-dimensional task-space ($n = 3$ for position management). The forward kinematics is a surjective function of the angles in joint-space.

$$\mathbf{x}(t) = f(\boldsymbol{\theta}(t)), \tag{1.1}$$

where $\mathbf{x}(t) \in \mathbb{R}^n$ and $\boldsymbol{\theta}(t) \in \mathbb{R}^m$ are the task-space and joint-space coordinates respectively. Remember, for a redundant articulated robot, $m > n$. Using the mechanical architecture and Denavit–Hartenberg (DH) parameters for a given articulated robot, forward kinematic mapping $f(.)$ is a nonlinear vector-valued function, which is straightforward to calculate. Nevertheless, instead of joint-space, the task for a robot is normally defined in the cartesian task-space. We are, therefore, more concerned in inverse mapping, i.e., mapping from the task-space to the joint-space. Using (1.1) inverse kinematics can be written as

$$\boldsymbol{\theta}(t) = f^{-1}(\mathbf{x}(t)), \tag{1.2}$$

Where $f^{-1}(.)$ represents the inverse kinematic function. Consider the $\mathbf{x}r(t)$ reference trajectory for end-effector location in cartesian task-space. The corresponding joint-space angles $\boldsymbol{\theta}_r(t)$, will fulfill the following relation, to trace this trajectory,

$$\mathbf{x}_r(t) = f(\boldsymbol{\theta}_r(t)). \tag{1.3}$$

Our goal is to solve this equation for the value of $\boldsymbol{\theta}_r(t)$. If there is a closed-form expression for $f^{-1}(.)$, we can solve the equation easily using $\boldsymbol{\theta}_r(t) = f^{-1}(\mathbf{x}_r(t))$. However, the forward kinematic mapping $f(.)$ is surjective-only, i.e., there are infinite solutions $\boldsymbol{\theta}r(t)$ in the joint space, which get mapped to $\mathbf{x}_r(t)$.

To solve the redundancy, i.e., to determine an optimum joint-space path out of infinite possible trajectories; we model the tracking control as follows:

$$\min_{\boldsymbol{\theta}(t)} \quad g_{tr}(\mathbf{x}_r(t), \boldsymbol{\theta}(t)), \tag{1.4}$$

where $g_{tr}(.)$ is the tracking objective function and defined as

$$g_{tr}(\mathbf{x}_r, \boldsymbol{\theta}) = ||\mathbf{x}_r - f(\boldsymbol{\theta})||_2. \tag{1.5}$$

where $\mathbf{x}_r$ is the current point on the reference trajectory and $\boldsymbol{\theta}$ are the current joint-angles.

Remark 1.1 In the formulation of the objective function, only the kinematic model of the articulated robot is used. As shown by recently published literature [6, 34], the kinematic control is intensively studied for the management of articulated robots. Besides academic research, kinematic management is also used extensively in commercial robotic systems such as ping-pong articulated robot [73], Adept Quattro 650HS [74], ABB IRB 360 [75], DOBOT, and UR 10 articulated robot.

1.2.2 Formulation of Obstacle Avoidance Penalty-Term

The numerical solution to problem (1.4) does not ensure collision avoidance. The strategy for avoiding obstacles is based on the principle: maximizing the minimum distance between the articulated robot and the obstacle. To integrate this principle into our optimization problem, we will formulate a second term for the objective function that penalizes the joint-space angles that bring the robot closer to an obstacle. The problem of obstacles avoidance is mathematically described as

$$\min_{\boldsymbol{\theta}(t)} \quad g_{OI}(O, \boldsymbol{\theta}(t)), \tag{1.6}$$

where $g_{OI}(.)$ is called obstacle avoidance penalty function; which is a function of $O \in \mathbb{R}^{no \times 3}$, 3D geometry of obstacle, i.e., cartesian coordinates of all its vertices, and θ, articulated robot joint angles. Here nO represents the total number of vertices in the obstacle's 3D model. The $g_{OI}(.)$ is defined as

$$g_{OI}(O, \theta) = \frac{1}{[\min_{i \in \{1,2,...,m\}}\{d_i(O, \theta)\}]^{\beta}}, \tag{1.7}$$

where θ denotes the current value joint-angles, m is the total number of joints in the articulated robot and $d_i(O, \theta)$ is the distance of ith joint from the O obstacle. The inversely proportional formulation means that the reduction in objective function value decreases the distance between the links and the barrier. β is a hyperparameter, and we found that $\beta = 1$ provides the best results from empirical observations during simulations. The distance value is determined using the GJK algorithm (for more information refer to Sect. 1.3.3)

$$d_i(O, \theta) = \text{GJK}(O, \mathcal{M}_i(\theta)), \tag{1.8}$$

here $i \in \{1, 2, \ldots, m\}$. Because GJK algorithm requires 3D-geometry of the both the obstacle and robot, we defined a function $\mathcal{M}_i(\theta) \in \mathbb{R}^{n_i \times 3}$ which returns the vertices of ith link. Similar to nO, n_i is the number of vertices within the 3D-geometry of ith link. It should be remembered that when robot moves, the position of the vertices varies, i.e., it is a function of θ joint-angles. The initial geometry, $\mathcal{M}i(0)$, is given by the manufacturer of the robot. The subsequent $\mathcal{M}(\theta)$ values are determined using

$$\mathcal{M}_i(\theta) = \mathcal{R}_i(\theta)\mathcal{M}_i(0) + \mathcal{T}_i(\theta)$$

where $\mathcal{R}_i(\theta)$ and $\mathcal{T}_i(\theta)$ are the rotation and translation matrix for the ith link. These matrices depends on kinematic model of articulated robot.

1.2.3 Constraints on Joint-Angles

A numerical solution to both problem (1.4) and (1.6) does not ensure that the final joint-space trajectory will lie within the mechanical limit of the robot. To ensure that the control action generated by the controller does not violate the joint-angle limits, the following constraint must be satisfied

$$\theta^- < \theta(t) < \theta^+, \tag{1.9}$$

where θ^- and θ^+ are the lower and upper limits on the joints-angles respectively. The value of these limits depend on the mechanical structure of the robot and the type of motors used to move the joints.

1.2.4 Combined Optimization Problem

Above we formulated three component of the final optimization problem: tracking (1.4), obstacle avoidance (1.6) and joint-angle limits (1.9). These can be combined into the following optimization problem

$$\min_{\theta(t)} \; g(O, \mathbf{x}_r(t), \theta(t))$$
$$\text{s.t.} \;\; \theta^- < \theta(t) < \theta^+, \tag{1.10}$$

where $g(.)$ is the unified objective function defined as

$$g(O, \mathbf{x}_r, \theta) = g_{tr}(\mathbf{x}_r, \theta) + \Lambda g_{OI}(O, \theta), \tag{1.11}$$

where Λ is a constant hyperparameter used to regulate the contribution of the individual term in the objective function. It controls the ratio of the optimizer's effort between tracking performance and obstacle avoidance. A value of $\Lambda = 0$ removes the contribution of the obstacle avoidance term. The value of Λ greatly affect obstacle avoidance performance. Its effect is discussed in detail in Sect. 1.4.

Although the penalty term approach rewards the optimizer to avoid the obstacle, however, in certain scenarios, the placement of obstacle makes it impossible to avoid it while following the reference trajectory; to avoid the collision in such a scenario, we add an inequality constraint to (1.10),

$$\min_{\theta(t)} \; g(O, \mathbf{x}_r(t), \theta(t))$$
$$\text{s.t.} \;\; \theta^- < \theta(t) < \theta^+$$
$$d_i(O, \theta(t)) > d_{min} \text{ for } i \in \{1, 2, \ldots, m\}. \tag{1.12}$$

The second constraint puts a hard lower-bound, d_{min}, on robot-obstacle distance.

Based on the above formulation, the complete form of the optimization problem can be written as

$$\min_{\theta(t)} \; ||\mathbf{x}_r(t) - f(\theta(t))||_2 \; + \Lambda \frac{1}{[\min_{i \in \{1,2,\ldots,m\}} \{d_i(O, \theta(t))\}]^{\beta}}$$
$$\text{s.t.} \;\; \theta^- < \theta(t) < \theta^+$$
$$d_i(O, \theta(t)) > d_{min} \;\; \text{for } i \in \{1, 2, \ldots, m\}. \tag{1.13}$$

The solution to this optimization problem gives the joint-space trajectory $\theta_r(t)$.

Fig. 1.1 The reward
generation mechanism used
to incorporate the penalty
term in objective function
(1.11)

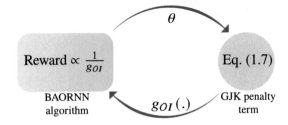

The obstacle avoidance objective function $g_{OI}(O, \boldsymbol{\theta})$ acts as a penalty term in the unified objective function above. When the articulated robot is moved far from the obstacle, the value of the penalty term becomes small, and the algorithm rewards the optimizer by reducing the overall value of the objective function. Figure 1.1 illustrates the award generation mechanism.

! Multiple Obstacles

In this chapter, specifically the case of a single obstacle is considered for the formulation of the optimization problem. However, a trivial approach to extend the formulation to the case of multiple obstacles is to modify the objective function as follow

$$g(O, \mathbf{x}_r, \boldsymbol{\theta}) = g_{tr}(\mathbf{x}_r, \boldsymbol{\theta}) + \Lambda \max_{i=1}^{k} g_{OI}(O_i, \boldsymbol{\theta}), \qquad (1.14)$$

where k denotes the total number of obstacles.

1.3 Design and Analysis of Management Algorithm

In this section, we will present the mathematical formulation of the BAORNN algorithm. Then we will briefly describe the GJK algorithm used for calculating the distance between the robot and5 the obstacles.

1.3.1 Algorithm Formulation

Following the problem formulation in Sect. 1.2, kinematic control , and obstacle avoidance effectively become equivalent to solving the (1.13) optimization problem in real-time while the articulated robot is moving. BAORNN algorithm imitates a beetle's behavior, which uses its pair of antennas and olfactory senses to examine an unfamiliar area in search of food (i.e., search for a region with maximum odor). Beetle tests the degree of odor on both antennas at each step before determining the direction of its next step. In particular, note the intermediate action, i.e., instead of going randomly in any direction, it uses only the olfactory sense to develop better intuition about the direction of the target and only then makes a strategic decision to take the next step. This overall behavior inspired us to integrate the idea of "visual robots" (analogous to the olfactory sense of antennae) into our management framework and develop a heuristic mechanism for managing the articulated robot.

Consider the articulated robot, at time-step k starts with joint-angles of $\boldsymbol{\theta}_k$. The algorithm imitates the direction of antennae by generating a normally distributed random vector $\mathbf{b} \in \mathbb{R}^m$. Using the vector $\mathbf{b}$, the position of antennae end-point can be calculated as

$$\boldsymbol{\theta}_{kL} = \boldsymbol{\theta}_k + \lambda_k \mathbf{b}, \qquad \boldsymbol{\theta}_{kR} = \boldsymbol{\theta}_k - \lambda_k \mathbf{b}, \qquad (1.15)$$

where λ_k is a hyperparameter denoting the antennae length, $\boldsymbol{\theta}_{kL}$ and $\boldsymbol{\theta}_{kR}$ denotes the location of left and right antennae respectively at time-step k. These vectors represent the direction of the next possible joint-angles. However, these vectors might not

---

Algorithm 1: BAORNN algorithm - Tracking & Obstacle avoidance

---

Input: kinematic model $f(.)$ and 3D-geometry matrix $\mathcal{M}_i(0)$ ($i \in \{1, 2, ..m\}$) of the articulated robot, 3D-geometery of the obstacle O, reference trajectory $\mathbf{x}_r(t) \in \mathbb{R}^n$, an objective function $g(O, \mathbf{x}, \boldsymbol{\theta})$. Additionally, the values of hyper-parameters: β, Λ, c_1 and c_2.

Output: An optimal trajectory $\boldsymbol{\theta}_r(t)$ in joint-space.

$\boldsymbol{\theta}_0 \leftarrow$ Initial joint coordinates

$k \leftarrow 0$ $k_{stop} \leftarrow$ maximum number of time-steps allowed

while $k < k_{stop}$ **do**

> Generate a normalized random direction vectors, $\mathbf{b} \in \mathbb{R}^m$ in the joint-space.
> Use the generated random vector to calculate the location of left and right antennae, $\boldsymbol{\theta}_{kL}$ and $\boldsymbol{\theta}_{kR}$ respectively, using (1.16).
> Project the location of these antennae on the constrained set Ω using the projection function $\mathcal{P}_\Omega$ as defined in (1.17) to . Calculate the value of objective function at both location using "Virtual robots" as defined in (1.18).
> Calculate he updated location in joint-space using (1.19).
> Check if the updated location improves the value of objective function using (1.21).
> Move the robot's angles to $\boldsymbol{\theta}_{k+1}$ and update the value of g_{k+1}.
> $k \leftarrow k+1$

end

---

satisfy the mechanical and obstacle avoidance constraints as outlined in the problem (1.13). Therefore, we project these vectors onto the constrained set

$$^{\Omega}\boldsymbol{\theta}_{kX} = \mathcal{P}_{\Omega}(\boldsymbol{\theta}_{kX}), \tag{1.16}$$

where $X \in \{L, R\}$, $\mathcal{P}_{\Omega}(.)$ is the projection function which confines the input inside the constrained set Ω. The set Ω is defined as,

$$\Omega = \{\boldsymbol{\theta} \in \mathbb{R}^m | \boldsymbol{\theta}^- < \boldsymbol{\theta} < \boldsymbol{\theta}^+ \wedge d_i(O, \boldsymbol{\theta}) > d_{min}\}.$$

There are several way to project a vector $\boldsymbol{\theta}$ on a set Ω. In this work, we chosse a computationally trivial approach and define the projection function as follow

$$\mathcal{P}_{\Omega}(\boldsymbol{\theta}_{kX}) = \begin{cases} \max\{\boldsymbol{\theta}^-, \min\{\boldsymbol{\theta}_{kX}, \boldsymbol{\theta}^+\}\} & \text{if } d_i > d_{min} \\ \boldsymbol{\theta}_k & \text{if } d_i < d_{min}, \end{cases} \tag{1.17}$$

where again $X \in \{L, R\}$, d_i is same as defined in (1.8). The value of objective function is then evaluated at the projected antennae locations $^{\Omega}\boldsymbol{\theta}_{kL}$ and $^{\Omega}\boldsymbol{\theta}_{kR}$ respectively

$$g_{kX} = g(O, \mathbf{x}_r(t), ^{\Omega}\boldsymbol{\theta}_{kX}), \tag{1.18}$$

where g_{kX} ($X \in \{L, R\}$) is the value of objective function at antenna locations.

We then use the value of the objective function calculated above to move the joint-angles in a direction where the value of the objective function is decreasing. We achieve this by using the following update rule

$$^{\Omega}\boldsymbol{\theta}'_{k+1} = \mathcal{P}_{\Omega}(\boldsymbol{\theta}_k - \delta_k(\lambda_k)\text{sign}(g_{kL} - g_{kR})\mathbf{b}), \tag{1.19}$$

where $^{\Omega}\boldsymbol{\theta}'_{k+1}$ is the updated location joint-angles projected on set Ω. The term $\text{sign}(g_L - g_R)\mathbf{b}$ ensures that the next step is taken toward a direction where the value of objective function will be smaller. $\delta_k(\lambda_k)$, is a hyperparameter denoting the step-size. The step-size is a function of antennae length λ_k; there relationship is discussed later. After calculating $^{\Omega}\boldsymbol{\theta}'_{k+1}$, the objective function is re-evaluated

$$g'_{k+1} = g(O, \mathbf{x}_r(t), ^{\Omega}\boldsymbol{\theta}'_{k+1}), \tag{1.20}$$

this new value g'_{k+1} is compared to the value of objective function at previous time-step g_k. If there is any improvement (i.e., the new value is smaller), then the joint-angles are moved to $^{\Omega}\boldsymbol{\theta}'_{k+1}$; otherwise, they remain the same

$$\boldsymbol{\theta}_{k+1} = \begin{cases} ^{\Omega}\boldsymbol{\theta}'_{k+1} & \text{if } g'_{k+1} < g_k \\ \boldsymbol{\theta}_k & \text{if } g'_{k+1} \geq g_k. \end{cases} \tag{1.21}$$

Similarly, the value of g_{k+1} is assigned to use in next iteration

$$g_{k+1} = \begin{cases} g'_{k+1} & \text{if } g'_{k+1} < g_k \\ g_k & \text{if } g'_{k+1} \geq g_k. \end{cases} \tag{1.22}$$

After moving to θ_{k+1}, the iterative procedure is repeated for the next time-steps. The steps of the proposed algorithm are systematically presented in Algorithm 1.

The choice of hyperparameters λ_k and $\delta_k(\lambda_k)$, can affect the speed of convergence. We found that the following rules for the selection of hyper-parameters by empirical analysis. They provide a reasonable level of performance

$$\lambda_k = c_1\sqrt{g'_k} \tag{1.23}$$

$$\delta_k(\lambda_k) = c_2\lambda_k \tag{1.24}$$

where c_1 and c_2 are constants. The above rules control the step-size and antenna length, making them large when the end-effector is far from the goal position and extremely small when reached near the goal. The small step-size is necessary to prevent the overshooting of end-effector near goal position. For c_1 and c_2 needs to be manually tuned.

The RNN architecture of the proposed algorithm is shown in Fig. 1.2a. The formulated RNN two layers. A temporal-feedback exists between the output of the second layer to the input of first. RNN architecture has a total of $3m + 6$ neurons. The neurons, depicted as circles, implement the functionality of projection function $\mathcal{P}_\Omega(.)$. Other neurons, depicted as curved rectangular boxes, represent "virtual robots"', and their activation function is $f(.)$. Similarly, the neurons represented by curved boxes (in cyan) use objective function evaluation $g(.)$ as their activation function.

1.3.2   Theoretical Proofs

Theorem 1.1 *For the tracking and obstacle avoidance of a redundant articulated robot, starting from an initial joint-space angles θ_0; the joint-space trajectory $\theta_r(t)$ generated by BAORNN algorithm is stable, i.e.,*

$$g_{k+1} \leq g_k, \quad \forall \ k \geq 0, \tag{1.25}$$

the values of objective function are monotonically decreasing.

Proof See Lemma 1 of [64].   □

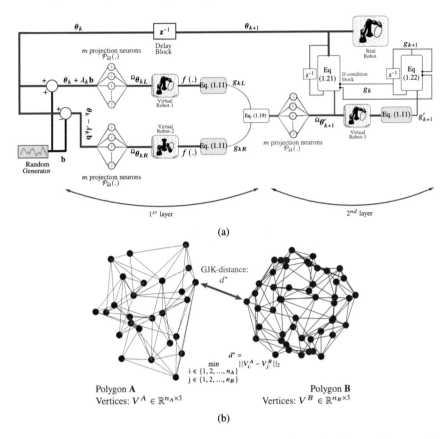

(a)

(b)

Fig. 1.2 a The topology of the RNN for the BAORNN algorithm. The diagram illustrates the working of the algorithm formulated in Sect. 1.3.1. **b** Illustration of GJK algorithm

Theorem 1.2 *For the tracking and obstacle avoidance of a redundant articulated robot, starting from an initial joint-space angles θ_0; the end-effector trajectory $f(\theta_r(t))$ is convergent to the reference trajectory $\mathbf{x}_r(t)$, i.e.,*

$$f(\theta(t)) \rightarrow \mathbf{x}_r(t), \quad as \quad t \rightarrow \infty. \tag{1.26}$$

Proof See Theorem 1 of [64].   □

1.3.3 GJK-Distance Algorithm

GJK algorithm is a numerically efficient algorithm, extensively using in computer graphics to calculate the minimum distance between two arbitrarily shaped convex

3D-polygons. Although, in case of obstacles and articulated robots, the 3D-geometry might be non-convex, however, the collision avoidance between their convex-hulls is a sufficient condition for the actual collision avoidance.

Consider two polygons A and B in 3D-space, the location of their vertices are defined by matrices $V^A \in \mathbb{R}^{n_A \times 3}$ and $V^B \in \mathbb{R}^{n_B \times 3}$ respectively. n_A and n_B are the numbers of vertices of polygon A and B, respectively. The GJK algorithm takes these matrices and calculates the distance between the closest vertices of the two polygons,

$$\text{GJK}(V^A, V^B) = \min_{\substack{i \in \{1, 2, \ldots, n_A\} \\ j \in \{1, 2, \ldots, n_B\}}} ||V_{i:}^A - V_{j:}^B||_2$$

where the notation $V_{i:}$ is used to represent the ith row of a matrix V. Figure 1.2b illustrates GJK-algorithm.

1.3.4   Computational Complexity

Now we present the theoretical analysis of computational complexity of the BAORNN algorithm. The first step in the algorithm is the generation of is a random vector $\mathbf{b}$ with m elements; the operation requires m floating-point operations. Next, we calculate θ_{kL} and θ_{kR}, each requiring m multiplication and m additions, totalling $4m$ floating-point operations. Next step requires the projection of two vectors using the projection function $f_{\Omega}(.)$, which require a total of $4m$ comparisons. Then we use (1.18) to calculate the value of objective function at both antennae location. The evaluation of objective function is the most computationally intensive step of the algorithm since it requires the calculation of Euclidean distance as well as GJK-distance, as given in (1.11). The calculation of Euclidean distance require a total of $3m - 1$ floating-point operations (m subtractions, m squaring operations and $m - 1$ additions). The calculation of GJK-distance depends on the number of vertices in the 3D models of two objects and require a total of $n_A + n_B$ operations, as shown by [76]. Where n_A and n_B are the numbers of vertices in the 3D model of both objects, respectively. For the case of robot and obstacle's distance, using the notation of Sect. 1.2.2, the total number of operation comes out to be $n_O \sum_{i=1}^{m} n_i$. Although this number is large, these operations are only required in the first iterations of the algorithm, the later iterations of GJK-algorithm are near-constant time, as pointed out by [76, 77]. Therefore, the total number of operations required by GJK-algorithm are effectively m. It means that a total of $4m + 2$ operations are required for evaluating the objective function; some additional operations are required for the scalar addition and multiplication as given in (1.11). Since objective function is evaluated twice in (1.18), therefore this step require a total of $8m + 4$ operations. The next step, as given in (1.19), requires a total of $2m + 1$ floating-point operations. Similarly, the subsequent step is again objective function evaluation requiring $4m + 2$ operations. The final step of the algorithm, as given in (1.21) and (1.22), require a total of $2m$ comparisons. Adding the floating-point operations required for each step as calculated above; the final count comes out to be $(m + 4m + 4m + (8m + 4) + (2m + 1) + (4m + 2) + 2m) = 25m + 7$.

This analysis demonstrates that the complexity of the proposed algorithm is $O(m)$, i.e., a linear function of the number of links. For the IIWA14 robot, a total of 182 operations are required in iterations. Modern embedded processors can efficiently perform floating-point operations of this order within a few hundred microseconds.

1.4 Simulation Methodology, Results and Discussion

Simulation methodology for evaluating the output of the proposed algorithm is provided in this section, along with the findings and discussion that have been obtained. KUKA LBR IIWA-14 model is used as a test-bench. The IIWA-14 has 7-DOF. The articulated robot 3D-model shown in Fig. 1.3.

1.4.1 Simulations

MATLAB Robotic System Toolbox [78] provides a simulation model of IIWA-14. The model represents an excellent depiction of the actual robot and acts as a desirable simulation test-bench. To test the obstacle avoidance performance, we placed an arbitrarily shaped obstacle in front of the robot. The simulation setup, including the articulated robot and obstacle, is shown in Fig. 1.3.

Two reference trajectories [36] were used in our simulations: a rectangular and a circular trajectory as shown in Fig. 1.4. The four vertices of the rectangular paths used in simulation are: $[0.2\ 0.6\ 0.8]^T$, $[-0.1\ 0.6\ 0.8]$, $[-0.1\ 0.6\ 0.2]$, and $[0.2\ 0.6\ 0.2]$. The total time for tracking the rectangular trajectory is 50 seconds. For generating the circular trajectory we used following equation

$$\mathbf{x}_r^{circle}(t) = \mathbf{C} + r \cos(2\pi t/T)\mathbf{A} + r \sin(2\pi t/T)\mathbf{B}. \tag{1.27}$$

where $\mathbf{C}$ denotes the vector to center of the circle, $\mathbf{A}$ and $\mathbf{B}$ are two perpendicular unit vectors defining the plane of the circle in 3D space, r is the radius of the circle. T is the total time to follow trace the trajectory. Following values were used in simulation: $\mathbf{C} = [0.0\ 0.6\ 0.5]$, $\mathbf{A} = [1\ 0\ 0]$, and $\mathbf{B} = [0\ 0\ 1]$. These values generate a circlular path in $x - z$ plane at $y = 0.6$. Without the loss of generality, the proposed algorithm works for an arbitrarily shaped reference-trajectory, given that the trajectories satisfy the mechanical consists of the articulated robot.

To methodically study the effect of the proposed algorithm, we removed the contribution of obstacle avoidance penalty term in the objective function, i.e., setting $\Lambda = 0$ in (1.13) and ignoring the 2nd constraint. Then we performed simulations with different values of Λ as discussed next.

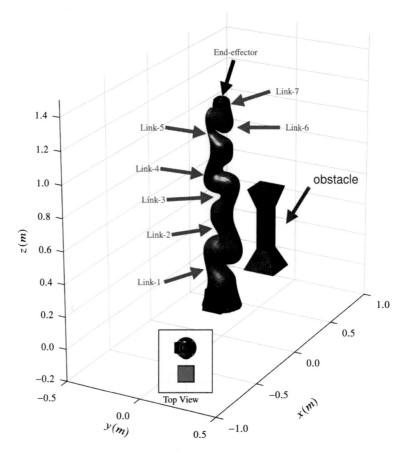

Fig. 1.3 3D model of KUKA LBR IIWA-14 7-DOF articulated robot with the obstacle used in simulations. The obstacle is placed in front of the operational region of the robot

1.4.2 Obstacle Avoidance Performance

The first batch of simulations analyzed the performance of the algorithm without the contribution of the obstacle avoidance term. The results of this scenario are shown in Fig. 1.4. It can be seen that in this case, the links of the articulated robot collide with the obstacle. This happens because the obstacle avoidance term has a weight of zero, causing the algorithm to ignore the obstacle effectively.

Next we conducted a set of experiments with varying values of Λ as defined in (1.11). Figures 1.5 and 1.6 show the response of the system is case of rectangular path. The The robot's joints are assumed to be start from home position, i.e., all joint-angles are zero at the beginning. Figure 1.5a–e summarizes the robot's response for $\Lambda = 0.002$. Figure 1.5a shows the motion of each links of the robot along with

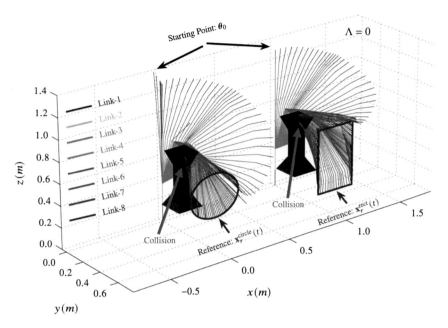

Fig. 1.4 Performance of tracking algorithm after switching-off the obstacle avoidance term, i.e., $\Lambda = 0$ as defined in (1.11). The links collide with obstacle for both trajectories

the reference path (shown in blue). Initially, the end-effector's trajectory lies far from the reference path because the starting from home configuration, the algorithm takes some time to converge to an optimal joint-space trajectory. Once the end-effector reaches near the reference path, it follows the path accurately. Top view of the robot's trajectory is also shown as inset graphic, which confirms that the links of the robot does not collide with the obstacle. Figure 1.5b shows the cartesian coordinates of end-effector motion and Fig. 1.5c shows the joint-space coordinates of the robot trajectory. It should be noted that the ripply response depicted in these trajectories is typical response of metaheuristic algorithms because of the stochastic nature. Figure 1.5d shows the position tracking error which is defined as $\mathbf{e}(t) = \mathbf{x}_r(t) - f(\boldsymbol{\theta}_r(t))$. Initially at $t = 0$, the value of tracking error is comparatively huge $\approx [0.5 \; -0.5 \; 0.7]^T$, however, after some time, the tracking error converges to zero. It also proves that the asymptomatic convergence of the controller, i.e., the tracking error converges to zero and does not rise again, except for some small ripples caused by the stochastic nature of the algorithm. Similarly, Fig. 1.5e shows the minimum distance of any link of the robot from the obstacle as defined in (1.7). A high value is preferable because it reduces the probability of collision in case of uncertainty in robot's model or obstacle's position. We repeated the same set of simulations with $\Lambda = 0.0002$. Figure 1.6a–e summarizes the robot's response. The major difference between these two situations is the contributon of obstacle avoidance term to the

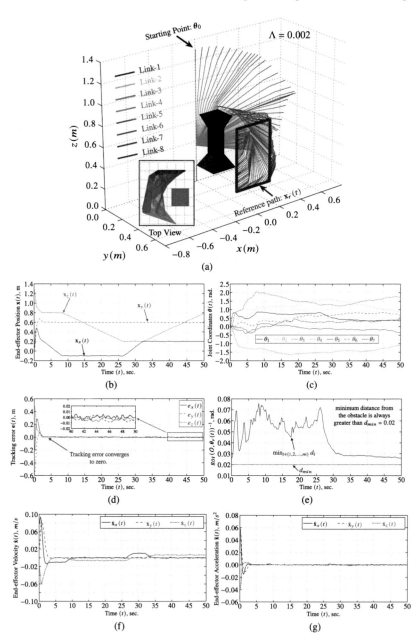

Fig. 1.5 Simulation results for rectangular trajectory tracking for values of $\Lambda = 0.002$ as defined in (1.11). **a** The trajectory of each link of articulated robot along with reference path. **b** Profile of task-space trajectory of the end-effector. **c** Profile of joint-space trajectory of the articulated robot. **d** Profile of the position tracking error. **e** Minimum GJK-distance of the robot from obstacle as defined in (1.7). **f** and **g** show the velocity and acceleration profiles respectively

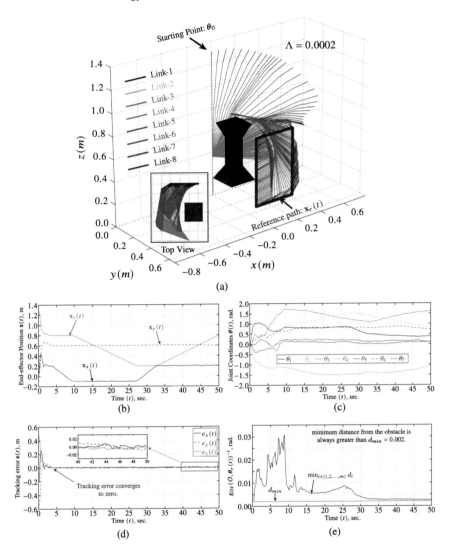

Fig. 1.6 Simulation results for rectangular trajectory tracking for values of $\Lambda = 0.0002$ as defined in (1.11). **a** The trajectory of each robot's link, along with reference path. **b** Profile of task-space trajectory of the end-effector. **c** Profile of joint-space trajectory of the articulated robot. **d** Profile of the position tracking error. **e** Minimum GJK-distance of the robot from obstacle as defined in (1.7). It must be noted that the minimum robot-obstacle distance for $\Lambda = 0.002$ as shown in Fig. 1.5 is much better (i.e., larger) as compared to $\Lambda = 0.0002$

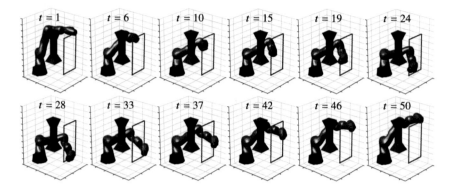

Fig. 1.7 Snapshots of the simulated model of the LBR IIWA-14, at several time-instants, while tracking the rectangular reference path for $\Lambda = 0.002$

final value of objective function. Figure 1.6e shows that the minimum robot-obstacle distance is smaller as compared to Fig. 1.5e, i.e., the links of robot were closer to the obstacle as compared to the latter case, increases the risk of collision. The same conclusion can be drawn from the inset graphics of Fig. 1.6a which shows that the links are much closer to obstacle in second case as compared to the first case. We had to reduce the value of d_{min} to 0.002 to successfully simulate a complete rectangular trajectory without colliding with an obstacle (Fig. 1.7).

The results for the trajectory tracking for the case of the circular path are illustrated in Figs. 1.8 and 1.9. The results are similar to that of the rectangular reference path. For the case when Λ is small, the overall distance between robot and obstacle decreases. However, if the value of Λ is large, the algorithm maintains a larger distance between obstacles and robots. However, it is worth mentioning that increasing the value too much will significantly degrade the performance of the tracking control because the algorithm will aggressively try to avoid the obstacle (Fig. 1.10).

1.4.3 Multiple Obstacle

We also performed a set of experiments using the multiple obstacles scenario, as shown in Fig. 1.11. The two obstacles were placed in front of the robot, and a circular path was given as a reference path. As shown in Fig. 1.11, if we don't consider the obstacle penalty-term in the objective function, i.e., $\Lambda = 0$ in (1.14), then the end-effector path intersect both of the obstacles. It will create a collision between the robot and an obstacle. However, if we increase the value of Λ to 0.002, then the penalty-term starts to influence the value of the objective function and force the management algorithm to calculate a joint-space trajectory, which avoids the obstacles. The result for $\Lambda = 0.005$ are shown in Fig. 1.12. Figure 1.12a shows the 3D path of robot's links. The difference between 3D paths shown in Figs. 1.11 and 1.12a clearly show

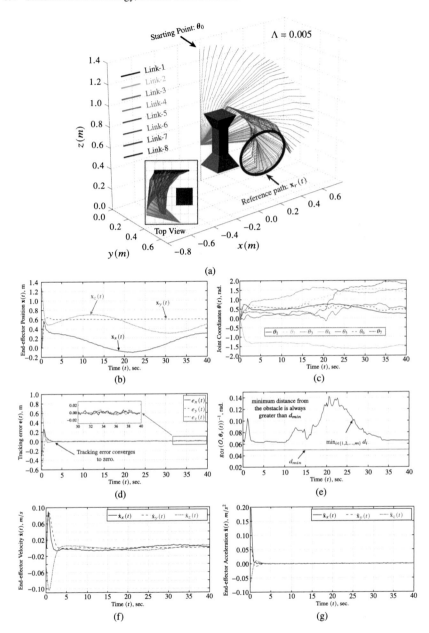

Fig. 1.8 Simulation results for circular trajectory tracking for values of $\Lambda = 0.005$ as defined in (1.11). **a** The trajectory of each robot's link along with reference path. **b** Profile of task-space trajectory of the end-effector. **c** Profile of joint-space trajectory of the articulated robot. **d** Profile of the position tracking error. **e** Minimum GJK-distance of the robot from obstacle as defined in (1.7). **f** and **g** shows the velocity and acceleration profile of the end-effector

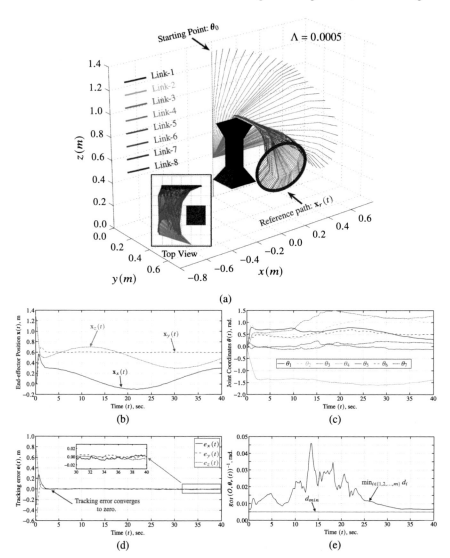

(a)

(b)

(c)

(d)

(e)

Fig. 1.9 Simulation results for circular trajectory tracking for values of $\Lambda = 0.0005$ as defined in (1.11). **a** The trajectory of each robot's link along with reference path. **b** Profile of task-space trajectory of the end-effector. **c** Profile of joint-space trajectory of the articulated robot. **d** Profile of the position tracking error. **e** Minimum GJK-distance of the robot from obstacle as defined in (1.7). It must be noted that the minimum robot-obstacle distance for $\Lambda = 0.005$, as shown in Fig. 1.8 is much better (i.e., larger) as compared to $\Lambda = 0.0005$

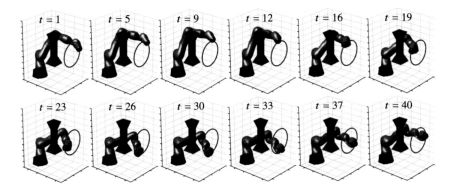

Fig. 1.10 Snapshots of the simulated model of the LBR IIWA-14, at several time-instants, while tracking the circular reference path for $\Lambda = 0.005$

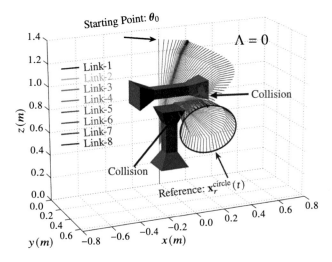

Fig. 1.11 Performance of tracking algorithm after switching-off the obstacle avoidance term, i.e., $\Lambda = 0$ as defined in (1.11). The links collide with obstacle for both trajectories

the contrast between two scenarios. Figure 1.12b shows the task-space coordinates of the end-effector. Figure 1.12c shows the profile of joint-angles. Figure 1.12d shows the profile of the tracking error, although it starts from a high value because the end-effector is located far from the reference path, however, as the management algorithm convergence, the tracking error becomes minimal. Figure 1.12e shows the minimum distance of any link of the robot from the obstacle. It can be seen that the value remains above 0.05. These results prove the efficacy of the proposed technique in case of multiple obstacles.

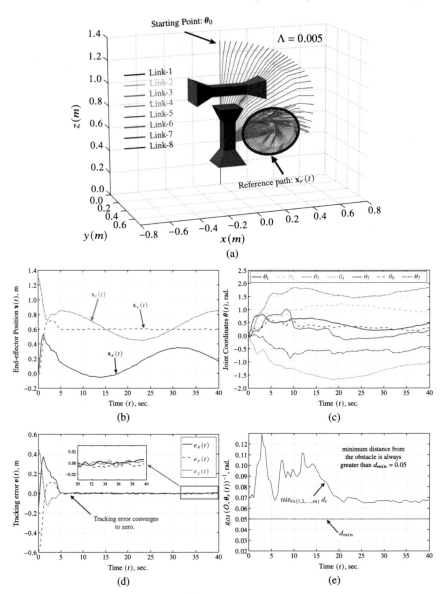

Fig. 1.12 Simulation results for rectangular trajectory tracking in case of multiple obstacle. $\Lambda = 0.005$ for this set of results. **a** The trajectory of each robot's link along with reference path. **b** Profile of task-space trajectory of the end-effector. **c** Profile of joint-space trajectory of the articulated robot. **d** Profile of the position tracking error. **e** Minimum GJK-distance of the robot from obstacle

1.5 Conclusion

In this chapter, we introduced a structure to address the problem of tracking and preventing obstacles in real-time at the same time. The suggested structure unifies the two objectives into a single constrained optimization problem. The penalty term method increases the efficiency of the proposed algorithm dramatically by deliberately rewarding the optimizer for avoiding the obstacles. This approach results in a joint-space control action which maximizes the distance between manipulator and obstacles. We proposed an RNN based on a metaheuristic optimization algorithm, called Beetle Antennae Olfactory, to solve the formulated optimization problem in real-time. A vital feature of the suggested structure is that it does not make assumptions on a specific shape of the obstacle. It uses the manipulator and obstacle's 3D-geometries to formulate the penalty term using GJK-algorithm. A potential application of such an approach involves the operation of a manipulator in a dynamically varying environment where the obstacle's shape varies in time. Applying the GJK-algorithm to measure manipulator-obstacle distance allows the algorithm to work independently of the shape of the robot or obstacle. Similarly, the proposed algorithm is also particularly useful for surgical robots, where it is critical to maintaining a safe distance to ensure safety between the manipulator's links and the patient. The theoretical treatment for proving the stability and convergence of the proposed algorithm is also provided. To prove the efficiency of the proposed algorithm, simulations using a 7-DOF industrial manipulator, KUKA LBR, are presented. Potential research direction to extend the current work includes reformulating the optimization problem to incorporate multiple obstacles while keeping the calculation of manipulator-manipulator distance computationally efficient.

References

1. C. Yang, C. Zeng, Y. Cong, N. Wang, and M. Wang, "A learning framework of adaptive manipulative skills from human to robot," *IEEE Trans. on Ind. Informatics*, vol. 15, no. 2, pp. 1153–1161, 2018.
2. H. M. La, T. H. Dinh, N. H. Pham, Q. P. Ha, and A. Q. Pham, "Automated robotic monitoring and inspection of steel structures and bridges," *Robotica*, vol. 37, no. 5, pp. 947–967, 2019.
3. C. Yang, Y. Jiang, Z. Li, W. He, and C.-Y. Su, "Neural control of bimanual robots with guaranteed global stability and motion precision," *IEEE Trans. on Ind. Informatics*, vol. 13, no. 3, pp. 1162–1171, 2016.
4. U. I. Khan and Z. Chen, "Natural oscillation gait in humanoid biped locomotion," *IEEE Transactions on Control Systems Technology*, 2019.
5. Z. Zhang, A. Beck, and N. Magnenat-Thalmann, "Human-like behavior generation based on head-arms model for robot tracking external targets and body parts," *IEEE Trans. on Cybern.*, vol. 45, no. 8, pp. 1390–1400, 2014.
6. Y. Zhang, S. Li, J. Zou, and A. H. Khan, "A passivity-based approach for kinematic control of redundant manipulators with constraints," *IEEE Trans. on Ind. Informatics*, 2019.
7. S. Li, J. He, Y. Li, and M. U. Rafique, "Distributed recurrent neural networks for cooperative control of manipulators: A game-theoretic perspective," *IEEE transactions on neural networks and learning systems*, vol. 28, no. 2, pp. 415–426, 2016.

8. S. Li, S. Chen, and B. Liu, "Accelerating a recurrent neural network to finite-time convergence for solving time-varying sylvester equation by using a sign-bi-power activation function," *Neural processing letters*, vol. 37, no. 2, pp. 189–205, 2013.
9. S. Li and Y. Li, "Nonlinearly activated neural network for solving time-varying complex sylvester equation," *IEEE Transactions on Cybernetics*, vol. 44, no. 8, pp. 1397–1407, 2013.
10. S. Li, Y. Zhang, and L. Jin, "Kinematic control of redundant manipulators using neural networks," *IEEE transactions on neural networks and learning systems*, vol. 28, no. 10, pp. 2243–2254, 2016.
11. K. Uzair and C. ZhiYong, "Natural gait analysis for a biped robot: jogging vs walking," *SCIENCE CHINA Information Sciences*.
12. C. Yang, G. Peng, L. Cheng, J. Na, and Z. Li, "Force sensorless admittance control for teleoperation of uncertain robot manipulator using neural networks," *IEEE Trans. on Syst., Man, and Cybern.: Syst.*, 2019.
13. H. M. La, R. Lim, and W. Sheng, "Multirobot cooperative learning for predator avoidance," *IEEE Trans. on Control Syst. Technology*, vol. 23, no. 1, pp. 52–63, 2014.
14. A. H. Khan, S. Li, and X. Luo, "Obstacle avoidance and tracking control of redundant robotic manipulator: An rnn based metaheuristic approach," *IEEE Transactions on Industrial Informatics*, 2019.
15. S. Li, Z.-H. You, H. Guo, X. Luo, and Z.-Q. Zhao, "Inverse-free extreme learning machine with optimal information updating," *IEEE transactions on cybernetics*, vol. 46, no. 5, pp. 1229–1241, 2015.
16. S. Li, B. Liu, and Y. Li, "Selective positive–negative feedback produces the winner-take-all competition in recurrent neural networks," *IEEE transactions on neural networks and learning systems*, vol. 24, no. 2, pp. 301–309, 2012.
17. L. Jin and S. Li, "Distributed task allocation of multiple robots: A control perspective," *IEEE Transactions on Systems, Man, and Cybernetics: Systems*, vol. 48, no. 5, pp. 693–701, 2016.
18. L. Jin, S. Li, H. M. La, and X. Luo, "Manipulability optimization of redundant manipulators using dynamic neural networks," *IEEE Transactions on Industrial Electronics*, vol. 64, no. 6, pp. 4710–4720, 2017.
19. L. Jin, S. Li, X. Luo, Y. Li, and B. Qin, "Neural dynamics for cooperative control of redundant robot manipulators," *IEEE Trans. on Ind. Informatics*, vol. 14, no. 9, pp. 3812–3821, 2018.
20. A. M. Zanchettin, L. Bascetta, and P. Rocco, "Achieving humanlike motion: Resolving redundancy for anthropomorphic ind. manipulators," *IEEE Robot. & Autom. Mag.*, vol. 20, no. 4, pp. 131–138, 2013.
21. D. Guo and Y. Zhang, "Acceleration-level inequality-based man scheme for obstacle avoidance of redundant robot manipulators," *IEEE Trans. on Ind. Electron.*, vol. 61, no. 12, pp. 6903–6914, 2014.
22. F. Basile, F. Caccavale, P. Chiacchio, J. Coppola, and C. Curatella, "Task-oriented motion planning for multi-arm robotic systems," *Robot. and Computer-Integrated Manuf.*, vol. 28, no. 5, pp. 569–582, 2012.
23. S. Li, Y. Guo, and B. Bingham, "Multi-robot cooperative control for monitoring and tracking dynamic plumes," in *2014 IEEE International Conference on Robotics and Automation (ICRA)*, pp. 67–73, IEEE, 2014.
24. S. Li, Y. Lou, and B. Liu, "Bluetooth aided mobile phone localization: a nonlinear neural circuit approach," *ACM Transactions on Embedded Computing Systems (TECS)*, vol. 13, no. 4, p. 78, 2014.
25. A. H. Khan, S. Li, X. Zhou, Y. Li, M. U. Khan, X. Luo, and H. Wang, "Neural & bio-inspired processing and robot control," *Frontiers in neurorobotics*, vol. 12, 2018.
26. Z. Zhang, S. Chen, X. Zhu, and Z. Yan, "Two hybrid end-effector posture-maintaining and obstacle-limits avoidance schemes for redundant robot manipulators," *IEEE Trans. on Ind. Informatics*, 2019.
27. G. Tevatia and S. Schaal, "Inverse kinematics for humanoid robots," in *Proceedings 2000 ICRA. Millennium Conf.. IEEE Intl. Conf. on Robot. and Autom. Symposia Proceedings (Cat. No. 00CH37065)*, vol. 1, pp. 294–299, IEEE, 2000.

28. A. Goldenberg, B. Benhabib, and R. Fenton, "A complete generalized solution to the inverse kinematics of robots," *IEEE Journ. on Robot. and Autom.*, vol. 1, no. 1, pp. 14–20, 1985.
29. Y.-J. Chen, M.-Y. Ju, and K.-S. Hwang, "A virtual torque-based approach to kinematic control of redundant manipulators," *IEEE Trans. on Ind. Electron.*, vol. 64, no. 2, pp. 1728–1736, 2016.
30. C.-S. Tsai, *Online Trajectory Generation for Robot Manipulators in Dynamic Environment–An Optimization-based Approach.* PhD thesis, UC Berkeley, 2014.
31. C. A. Klein and C.-H. Huang, "Review of pseudoinverse control for use with kinematically redundant manipulators," *IEEE Trans. on Syst., Man, and Cybern.*, no. 2, pp. 245–250, 1983.
32. J. Józwik, D. Ostrowski, P. Jarosz, and D. Mika, "Industrial robot repeatability testing with high speed camera phantom v2511," vol. 10, no. 32, 2016.
33. Y. M. Zhao, Y. Lin, F. Xi, and S. Guo, "Calibration-based iterative learning control for path tracking of industrial robots," *IEEE Trans on Ind. Electron.*, vol. 62, no. 5, pp. 2921–2929, 2014.
34. D. Chen, Y. Zhang, and S. Li, "Tracking control of robot manipulators with unknown models: A jacobian-matrix-adaption method," *IEEE Trans. on Ind. Informatics*, vol. 14, no. 7, pp. 3044–3053, 2017.
35. V. Lippiello, J. Cacace, A. Santamaria-Navarro, J. Andrade-Cetto, M. A. Trujillo, Y. R. Esteves, and A. Viguria, "Hybrid visual servoing with hierarchical task composition for aerial manipulation," *IEEE Robot. and Autom. Letters*, vol. 1, no. 1, pp. 259–266, 2015.
36. S. Li, S. Chen, B. Liu, Y. Li, and Y. Liang, "Decentralized kinematic control of a class of collaborative redundant manipulators via recurrent neural networks," *Neurocomputing*, vol. 91, pp. 1–10, 2012.
37. B. Liao and W. Liu, "Pseudoinverse-type bi-criteria minimization scheme for redundancy resolution of robot manipulators," *Robotica*, vol. 33, no. 10, pp. 2100–2113, 2015.
38. L. Xiao, S. Li, F.-J. Lin, Z. Tan, and A. H. Khan, "Zeroing neural dynamics for control design: comprehensive analysis on stability, robustness, and convergence speed," *IEEE Transactions on Industrial Informatics*, vol. 15, no. 5, pp. 2605–2616, 2018.
39. W. He, Z. Yan, Y. Sun, Y. Ou, and C. Sun, "Neural-learning-based control for a constrained robotic manipulator with flexible joints," *IEEE Trans. on neural networks and learning Syst.*, vol. 29, no. 12, pp. 5993–6003, 2018.
40. H. Wang and S. Kang, "Adaptive neural command filtered tracking control for flexible robotic manipulator with input dead-zone," *IEEE Access*, vol. 7, pp. 22675–22683, 2019.
41. H. Ding and S. K. Tso, "A fully neural-network-based planning scheme for torque minimization of redundant manipulators," *IEEE Trans. on Ind. Electron.*, vol. 46, no. 1, pp. 199–206, 1999.
42. W. He, Z. Yin, and C. Sun, "Adaptive neural network control of a marine vessel with constraints using the asymmetric barrier lyapunov function," *IEEE Trans. on Cybern.*, vol. 47, no. 7, pp. 1641–1651, 2016.
43. C. Yang, Y. Jiang, J. Na, Z. Li, L. Cheng, and C.-Y. Su, "Finite-time convergence adaptive fuzzy control for dual-arm robot with unknown kinematics and dynamics," vol. 27, no. 3, pp. 574–588, 2018.
44. J. Na, B. Jing, Y. Huang, G. Gao, and C. Zhang, "Unknown system dynamics estimator for motion control of nonlinear robotic systems," 2019.
45. H. Wang, Y. Zou, P. X. Liu, and X. Liu, "Robust fuzzy adaptive funnel control of nonlinear systems with dynamic uncertainties," vol. 314, pp. 299–309, 2018.
46. X. Jiang, S. Li, B. Luo, and Q. Meng, "Source exploration for an under-actuated system: A control-theoretic paradigm," *IEEE Transactions on Control Systems Technology*, 2019.
47. Y. Zhang, S. Li, and X. Jiang, "Near-optimal control without solving hjb equations and its applications," *IEEE Transactions on Industrial Electronics*, vol. 65, no. 9, pp. 7173–7184, 2018.
48. X. Jiang and S. Li, "Plume front tracking in unknown environments by estimation and control," *IEEE Transactions on Industrial Informatics*, vol. 15, no. 2, pp. 911–921, 2018.
49. A. T. Khan and S. Li, "A survey on blockchain technology and its potential applications in distributed control and cooperative robots," *arXiv preprint* arXiv:1812.05452, 2018.

50. U. A. Fiaz and J. S. Baras, "A hybrid compositional approach to optimal mission planning for multi-rotor uavs using metric temporal logic," *arXiv preprint* arXiv:1904.03830, 2019.
51. O. Khatib, "Real-time obstacle avoidance for manipulators and mobile robots," in *Autonomous robot vehicles*, pp. 396–404, Springer, 1986.
52. F. Flacco, T. Kröger, A. De Luca, and O. Khatib, "A depth space approach to human-robot collision avoidance," in *2012 IEEE Intl. Conf. on Robot. and Autom.*, pp. 338–345, IEEE, 2012.
53. X. Jiang and S. Li, "Beetle antennae search without parameter tuning (bas-wpt) for multi-objective optimization," *arXiv preprint* arXiv:1711.02395, 2017.
54. D. Chen, S. Li, F.-J. Lin, and Q. Wu, "New super-twisting zeroing neural-dynamics model for tracking control of parallel robots: A finite-time and robust solution," *IEEE transactions on cybernetics*, 2019.
55. D. Chen, S. Li, W. Li, and Q. Wu, "A multi-level simultaneous minimization scheme applied to jerk-bounded redundant robot manipulators," *IEEE Transactions on Automation Science and Engineering*, 2019.
56. D. Chen, Y. Zhang, and S. Li, "Tracking control of robot manipulators with unknown models: A jacobian-matrix-adaption method," *IEEE Transactions on Industrial Informatics*, vol. 14, no. 7, pp. 3044–3053, 2017.
57. E. G. Gilbert, D. W. Johnson, and S. S. Keerthi, "A fast procedure for computing the distance between complex objects in 3d space," *IEEE Journ. on Robot. and Autom.*, vol. 4, no. 2, pp. 193–203, 1988.
58. X.-S. Yang, *Nature-inspired metaheuristic algorithms*. Luniver, 2010.
59. A. H. Khan, Z. Shao, S. Li, Q. Wang, and N. Guan, "Which is the best pid variant for pneumatic soft robots? an experimental study," *IEEE/CAA Journal of Automatica Sinica*, vol. 6, no. 1, p. 1, 2019.
60. S. Li, Z. Wang, and Y. Li, "Using laplacian eigenmap as heuristic information to solve nonlinear constraints defined on a graph and its application in distributed range-free localization of wireless sensor networks," *Neural processing letters*, vol. 37, no. 3, pp. 411–424, 2013.
61. S. Li, R. Kong, and Y. Guo, "Cooperative distributed source seeking by multiple robots: Algorithms and experiments," *IEEE/ASME Transactions on mechatronics*, vol. 19, no. 6, pp. 1810–1820, 2014.
62. A. T. Khan, S. L. Senior, P. S. Stanimirovic, and Y. Zhang, "Model-free optimization using eagle perching optimizer," *arXiv preprint* arXiv:1807.02754, 2018.
63. X. Jiang and S. Li, "Bas: beetle antennae search algorithm for optimization problems," *arXiv preprint* arXiv:1710.10724, 2017.
64. Y. Zhang, S. Li, and B. Xu, "Convergence analysis of beetle antennae search algorithm and its applications," *arXiv preprint* arXiv:1904.02397, 2019.
65. A. H. Khan, X. Cao, S. Li, V. N. Katsikis, and L. Liao, "Bas-adam: An adam based approach to improve the performance of beetle antennae search optimizer," *IEEE/CAA Journal of Automatica Sinica*, vol. 7, no. 2, pp. 461–471, 2020.
66. A. H. Khan, S. Li, and X. Bin, "Bas-swarm: A nature-inspired metaheuristic algorithm with applications in machine learning," *Soft Computing*, vol. 1, no. 1, p. 1, 2019.
67. A. H. Khan, X. Cao, S. Li, and C. Luo, "Using social behavior of beetles to establish a computational model for operational management," *IEEE Transactions on Computational Social Systems*, vol. 7, no. 2, pp. 492–502, 2020.
68. Z. Zhu, Z. Zhang, W. Man, X. Tong, J. Qiu, and F. Li, "A new beetle antennae search algorithm for multi-objective energy management in microgrid," in *2018 13th IEEE Conf. on Ind. Electron. and Applications (ICIEA)*, pp. 1599–1603, IEEE, 2018.
69. X. Yin and Y. Ma, "Aggregation service function chain mapping plan based on beetle antennae search algorithm," in *Proceedings of the 2nd Intl. Conf. on Telecommunications and Communication Engineering*, pp. 225–230, ACM, 2018.
70. Q. Wu, X. Shen, Y. Jin, Z. Chen, S. Li, A. H. Khan, and D. Chen, "Intelligent beetle antennae search for uav sensing and avoidance of obstacles," *Sensors*, vol. 19, no. 8, p. 1758, 2019.
71. A. H. Khan and S. Li, "Tracking control of redundant manipulator under active remote center of motion constraints: An rnn-based metaheuristic approach," *SCIENCE CHINA Information Sciences*, 2019.

72. A. H. Khan, S. Li, D. Chen, and L. Liao, "Tracking control of redundant mobile manipulator: An rnn based metaheuristic approach," *Neurocomputing*, 2020.

73. S. Huang, Y. Peng, W. Wei, and J. Xiang, "Clamping weighted least-norm method for the manipulator kinematic control with constraints," *Intl. Journ. of Control*, vol. 89, no. 11, pp. 2240–2249, 2016.

74. G. Wu, "Kinematic analysis and optimal design of a wall-mounted four-limb parallel schönflies-motion robot for pick-and-place operations," *Journ. of Intelligent & Robotic Syst.*, vol. 85, no. 3-4, pp. 663–677, 2017.

75. I. Al-Naimi, A. Taeim, and N. Alajdah, "Fully-automated parallel-kinematic robot for multitask ind. operations," in *2018 15th Intl. Multi-Conf. on Syst., Signals & Devices*, pp. 390–395, IEEE, 2018.

76. M. Montanari, N. Petrinic, and E. Barbieri, "Improving the gjk algorithm for faster and more reliable distance queries between convex objects," *ACM Trans. on Graphics (TOG)*, vol. 36, no. 3, p. 30, 2017.

77. C. J. Ong and E. G. Gilbert, "The gilbert-johnson-keerthi distance algorithm: A fast version for incremental motions," in *Proceedings of Intl. Conf. on Robot. and Autom.*, vol. 2, pp. 1183–1189, IEEE, 1997.

78. P. I. Corke *et al.*, "A robotics toolbox for matlab," *IEEE Robot. & Autom. Mag.*, vol. 3, no. 1, pp. 24–32, 1996.

Chapter 2
Management of Soft Agents with Structural Uncertainty

Abstract As discussed in the preface of this brief, the articulated agents have two parts: the mechanical structure consisting of joints and links and an end-effector to manipulate the objects. Chapters two and three of this brief focuses explicitly on the second part, i.e., the end-effector of the articulated agent. Soft agents are extensively considered for replacing traditional rigid end-effectors. This chapter provides an experimental analysis of soft agents, also known as soft robots, that are structurally flexible. Soft agents are produced using soft materials and, in recent years, gained growing research attention both in academia and industry. Soft agents have an exciting feature of inherent safety in human interaction; therefore, they have become a center of attention for use in a robotic system where safe human interaction is needed. Nevertheless, the mathematical model of these soft agents is highly non-linear because of structural flexibility, and an infinite degree of freedom (DOF) is required to model their behavior accurately. For this reason, formulating a robust control strategy to control the position (or orientation) of these soft agents accurately and optimizing their dynamic behavior remains a challenging task. Model-free controllers, such as Proportional Integral Derivative (PID), are actually the most widely used soft agent control strategies. This chapter presents a comprehensive experimental analysis to thoroughly characterize the PID controller's behavior for the soft agents and to define their unique properties as opposed to rigid-robots. In addition, we also present a model-free parameter tuning technique to optimize the parameters of the PID controller, using the coordinate descent algorithm. In this chapter, we are studying the behavior of the PID controller's variants. For the case of manual tuning, we tested its performance using the Ziegler-Nichols method as well as automatic tuning, using coordinate descent. Statistically, the experimental results demonstrate the efficacy of the proposed automated tuning algorithm. Besides, we empirically demonstrated that the PID controller essentially reduces to the PI controller in the case of soft agents. This behavior was found in the case of manual and automatic parameter tuning experiments; a reason for eliminating the derivative term was also discussed here.

2.1 Introduction

Instead of conventional rigid structures, soft agents, i.e., robotic systems made from - soft materials (hereafter simplified as "soft agents"), are increasingly gaining attention. The use of soft structures opens up new possibilities to address the issues posed by classical rigid-robots; however, they also lead to new challenges. How to correctly model and monitor these systems is paramount among these challenges. A soft structure has an infinite degree of freedom, which makes it impossible to build a model as accurate as of that of a rigid structure [1–4]. This makes it challenging to regulate the motion of these soft robots, especially the tuning of their dynamic responses. In many applications, it raises serious issues, such as rehabilitation, where fine-grained control of muscles assisted by a soft structure is required. Another example is in high-speed applications such as industrial soft grippers; it is essential to fine-tune the dynamic responses.

Being a developing area of robotics, soft robotics, still have limited research work on the formulation of accurate modeling and dynamic control strategies. Vikas et al. [5, 6] proposed a graph-based, model-free framework for controlling the locomotion of soft agents. Calisti et al. [7, 8] found the inspiration from nature to formulate a control algorithm and proposed control strategies based on the motion of aquatic life. However, their algorithms only deal with the coarse-grained movement of the soft agents, instead of controlling the motion at a fine-grained level and tuning the dynamic response. Reymundo et al. [9] build a linear model using regression and used statistical data to estimate the model parameters. The assumption of linear response and the absence of a feedback loop can contribute to the instability of the soft robotic system. Frederick et al. [10–12] proposed control algorithm based on classical technique of Finite Element Method (FEM). FEM is potentially able to provide high accuracy performance but requires detailed knowledge of the mechanical parameters of the soft agent. Furthermore, FEM-based can not be executed in real-time because of high computation costs. One solution is to run the FEM analysis in a feedforward, open-loop manner. Nevertheless, this makes the control algorithm prone to model errors and can potentially render the system unstable. Marchese et al. [13–15] proposed a model-based control strategy to optimize the dynamic responses of soft agents.

Control algorithms based on approximate mathematical model of the soft agents have also been proposed [16–26]. Ni et al. [27] introduced an approach to managing the dynamic response of the soft agents by attaching mechanical damper to the body agents. Similarly, Wei et al. [28] and Li et al. [29] presented a design of soft actuators with particle chambers. The general principle behind these approaches is dissipating excessive mechanical energy by using external factors. However, the cost increment caused by new mechanical components and increased bulkiness makes these techniques less desirable. Luo et al. [30] proposed a Sliding-Mode Controller (SMC) based controller. However, their proposed controller requires careful manual tuning of the controller parameter. These strategies depend heavily on the analytical model of the soft agent and do not adapt to the mechanical variations. However, soft agents usually suffer from inevitable wear and tear [31] modifying the mathematical model; thereby, deteriorating their performance.

If the accurate mathematcal model or mechanical parameters of soft agent is not known a priori, model-free PID controllers become a desirable option for accurate control and dynamic response tuning [32, 33]. Several variants of the PID have been proposed in literature [34–41]. In this chapter, we conduct a comprehensive comparison of these variants. We revaluated the performance of each variants and highlight their key features [42–49]. The main contributions of this chapter can be summarized as:

- Comprehensive experimental study to analyze the comparative performance of the PID controllers and their ability to control soft agents. This chapter investigates three different variants of PID; regular PID, piecewise PID, and fuzzy PID. Two different types of parameter tuning algorithm are also discussed; Ziegler-Nichols for manual tuning and coordinate descent automatic tuning.
- This work uses extensive experiments and the resulting dynamic response data to demonstrate that the dynamical behaviors of soft agents are inherently different from that of rigid-robots.
- Following the identified inherent differences of PID control for soft and rigid-body robots, six types of dominating PID variants are systematically evaluated and compared on an experimental platform. The best PID variant is identified, and the rationale of its outstanding performance is also established.

The remainder of this chapter is organized as follows. Section 2.2 discusses literature review and related works. Section 2.3.1 presents the formulation of variants of PID-controller along with the description of parameter tuning algorithms. Section 2.4 describes the experimental platform, methodology, and discuss the results. Section 2.5 concludes this chapter.

2.2 Literature Survey

Recent research on soft agents is focused on their design, fabrication, control, applications. In this section, a brief review of recent advances in soft robotics is presented. Table 2.1 contains a summary of this review.

Soft agents are being promoted as counterparts to conventional rigid mechanical robots, while offering the equivalent level of functionality, with the inherent benefit of flexibility. For example, in conventional robotic systems, movement is produced using linear actuators such as linear DC motors. To provide the same functionality with soft structures, [50, 51] proposed a novel design of soft agent actuated using pneumatic systems. Likewise, to produce circular motions, designs of soft agents with the ability to bend on actuation are proposed in literature [52–56]. Similarly, designs of soft agents capable of three-dimensional motion have also been proposed [57].

Table 2.1 Summary of recent work in soft Robotics

	Design and fabrication	Sensing	Modelling	Control	Applications
Linear actuator [50, 51]	Yes	No	Static and Dynamic	No	No
Bending actuator [52–54]	Yes	No	Static	No	No
Other actuators [57, 60–62, 81]	Yes	No	Static	No	No
Linear sensing [63]	No	Yes	Static	No	No
Embedded sensing [64, 65]	Yes	Yes	Static	No	Yes[1]
Soft robotic systems [69, 74]	Yes	No	Static	Static	Yes[2]
model-free control [5, 6, 9]	No	No	Static	Static	No
Bio-inspired control [7, 8]	Yes	No	Static	Static	No
Finite Element Method (FEM) [10]	No	No	Static	Static	No
Dynamic control [13–15]	No	No	Static and dynamic	Static and dynamic	No

[1] Artificial skin, human gait measurement, wearable assistant robot etc.
[2] Hand, ankle-foot and shoulder rehabilitation, assistant system, soft variable length gripper etc

One of the primary contrasts between traditional rigid robots and soft agents is the calculation of a deterministically precise mathematical model. Traditional rigid robots can be precisely modeled because their motion is deterministic and accurately predictable. However, soft agents are flexible, theoretically requiring an infinite number of parameters to model them accurately [58]. Lack of an accurate mathematical model makes it challenging to create an accurate control algorithm. Estimation based techniques have been explored to model the soft agent as a linear system [59]; however, those estimations fail to model the nonlinear dynamics.

Soft agents use a different mode of actuation as compared to rigid robots. Most soft agents use a pneumatic system for actuation. Although pneumatics remains an excellent option for the actuation of soft structures, however, their bulkiness, weight, less portablity, and control delays remain an issue. Alternative actuation mechanisms have also been proposed. For example, the Dielectric Elastomer Actuators (DEA) [60–62], have been used to make soft agents. They are capable of producing large actuation when an electric voltage is applied across them.

Because of their soft construction, soft agents require unconventional sensing mechanisms to measure their motion as compared to rigid robots. The compliant nature of soft agents makes it possible to combine sensory mechanisms as part of the fabrication material. Felt et al. [63] proposed using an electrically conductive sleeve to encapsulate the body of the soft actuator. The motion of the robot is estimated using the variation in the electrical properties of the sleeve. Attempts have also been made to embed the sensor inside the body of soft agent [64–68].

Most of the literature on soft robots is centered around their rehabilitation applications [9, 69–73]. Traditional rehabilitation robots use rigid exoskeletons to assist the patient in exercises. Soft agents are more suitable in such applications because of their flexibility, compliance, and human-skin like appearance, which help in quickening the recovery. Soft grippers [74] have also shown to be competitive with rigid grippers. These grippers can easily handle delicate objects of various sizes and shapes. Due to this reason, they are widely being used as end-effector on industrial articulated agents [45, 46, 75–80].

2.3 PID-Controllers for Soft Agents

In this section, mathematical formulation of three such PID variant controllers is presented. The variants presented here include Piecewise PID and Fuzzy PID, whcih differ in the way PID parameters adapt with input reference angle. This section also present automatic algorithm to obtain optimal controller parametrs by optimizing the dynamic response of a soft agent.

2.3.1 Variants of PID Controllers

The following presents the formulation of a regular PID, a piecewise PID, and a fuzzy PID, which are used for comparison in this chapter, and discuss their differences.

2.3.1.1 A Regular PID-Controller

Let us denote a state of the soft agent as $\theta(t)$, and the desired or reference state as θ_r. The system error is then $e(t) = \theta_r - \theta(t)$, and the system control input is denoted as u(t). Based on this notation, a regular PID-controller defines the system input $u(t)$ as

$$u(t) = K_p e(t) + K_i \int_0^t e(\tau)d\tau + K_d \frac{d}{dt}e(t), \tag{2.1}$$

where K_p, K_i, and K_d are controller parameters. These controller parameters can be adjusted to tune the dynamic responses of the system. The above formulation assumes a continuous time system. In case sensor data acquisition and actuation take place in discrete time, (2.1) becomes

$$u[n] = K_p e[n] + K_i \sum_{i=1}^{n} e[i]\Delta t_i + K_d \frac{\Delta e_n}{\Delta t_n}, \tag{2.2}$$

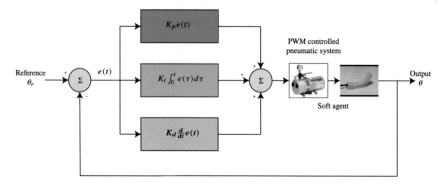

Fig. 2.1 Schematic diagram of canonical form of PID-controller for soft muscle control

where n is the discrete index number, $e[n] = \theta_r - \theta[n]$, $\Delta e_n = e[n] - e[n-1]$, and Δt_n is the time difference between two consecutive sensor readings, i.e. ($\theta[n]$ and $\theta[n-1]$). For simplicity of notation, let us denote the parameter vector and system error vector as

$$\mathbb{K} = \begin{bmatrix} K_p \\ K_i \\ K_d \end{bmatrix}, \qquad \mathbb{E}[n] = \begin{bmatrix} e[n] \\ \sum_{i=1}^{n} e[i]\Delta t_i \\ \Delta e_n / \Delta t_n \end{bmatrix}.$$

Using this, Eq. (2.2) becomes:

$$u[n] = \mathbb{K}^T \mathbb{E}[n]. \tag{2.3}$$

The schematic diagram of the PID-controller used in our experimental system is shown in Fig. 2.1. The main task in PID control is to adjust vector $\mathbb{K}$ to obtain the desired dynamic responses. The algorithms to tune $\mathbb{K}$ will be explained later.

2.3.1.2 Piecewise PID-Controller

The regular PID-controller of Eqs. (2.2) and (2.3) fixes the control parameter vector $\mathbb{K}$ for all possible values of reference input θ_r. However, to fine tune dynamic responses, different θ_rs needs different optimal control parameters. In other words, the optimal $\mathbb{K}$ is dependent on the given θ_r. A piecewise PID matches this demand by expressing $\mathbb{K}$ as a piecewise constant function of θ_r. The valid range of θ_r is divided into several subranges and the value of $\mathbb{K}$ is separately tuned for each subrange. That is, $\mathbb{K}$ now becomes $\mathbb{K}(\theta_r)$. Correspondingly, Formula (2.3) becomes

$$u[n] = \mathbb{K}(\theta_r)^T \mathbb{E}[n]. \tag{2.4}$$

Figure 2.2 illustrates the concept of the piecewise PID control parameter vector $\mathbb{K}$.

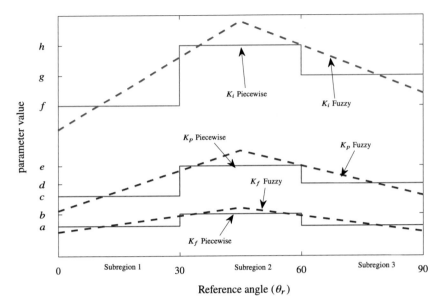

Fig. 2.2 Parameter variation in a Piecewise PID (solid lines) versus a Fuzzy PID (dashed lines)

2.3.1.3 Fuzzy PID-Controller

Although a piecewise PID increases flexibility in tuning control parameter vector $\mathbb{K}$, abrupt changes still occur at the boundaries of the subranges (see Fig. 2.2). Fuzzy PID further increases the flexibility if the controller parameters $\mathbb{K}$. Instead of defining $\mathbb{K}(\theta_r)$ as a piecewise constant function, it is defined $\mathbb{K}(\theta_r)$ as a piecewise linear and continuous function of θ_r. This formulation allows for a continuous change in the control parameter vector $\mathbb{K}$ at the boundaries of the subranges. The concept of the fuzzy control parameter vector is shown in Fig. 2.2. The mathematical formulation of the Fuzzy PID-controller is the same as in Eq. (2.4); the only difference is how $\mathbb{K}$ changes with θ_r.

2.3.2 Selection of Optimal Parameters

After formulation, the second task in implementing the PID-controllers as mentioned earlier is to tune the control parameter vector $\mathbb{K}$, for fine-grained control or optimal dynamic responses. Two methods to tune $\mathbb{K}$ will be discussed. The first is the manual method, which involves visually observing the performance of the system with different values of $\mathbb{K}$ and selecting an optimal value. The second method is to tune $\mathbb{K}$ using algorithms automatically. Both of these tuning methods are explained below.

? Why only the model-free control algorithms are considered?

Generally, the controllers can be classified into two types; model-based and model-free. Model-based algorithms require an accurate formulation of the mathematical model of the controlled system. However, the mathematical model soft agents are highly nonlinear systems, and estimating a linear model can only be used to approximate them in a small vicinity of the estimating point. For example, if the model is estimated at $\theta = 0$, then it will show good accuracy in a small neighborhood but give significant modeling errors at farther values. Since PID is basically a model-free controller, therefore, we directly rely on the output of the soft agent to tune the PID parameters instead of using an estimated mathematical model. System identification has also been studied in the soft robotic literature [59], which use the data collected from soft agent to estimate a second-order linear mathematical model using least-square estimation approach. In our current work, the objective is to present a controller formulation completely independent of the agent's model.

2.3.2.1 Manual Tuning—Ziegler-Nichols method

Manual tuning involves adjusting the control parameter vector $\mathbb{K}$ by hand and observing the corresponding system performances. If the performances are undesirable, then $\mathbb{K}$ should be changed manually, and the performances observed again, and so on. In this way, by manually adjusting $\mathbb{K}$ in a hit-and-miss approach, it is possible to achieve the desired system performances. This approach labor-intensive and depends heavily on the experience and judgment of the operators.

We used the Ziegler-Nichols method to manually tune the PID parameters which goes as follow: given θ_r, we first set K_i and K_d to zero and increase K_p until the step response of the control system is in converging oscillations. If the steady-state response contains a constant error, then K_p should be kept constant and K_i increased until the steady-state response error decreases to zero. At this stage, we can further fine-tune the oscillations behavior of the step response: settling time and overshoot. For example, to reduce overshooting, K_i should be kept constant, and K_d should be increased.

2.3.2.2 Automatic Tuning—Coordinate Descent

In contrast to manual tuning, automatic tuning is more principled and is based on rigorous algorithms. Several metaheuristic optimization algorithms have been proposed in the literature to tune the controller parameter [82–84]. However, most of these algorithms assume that the mathematical model of the controlled system is known apriori. However, in the case of soft agents, such assumptions are not practical. Therefore, we need to rely on the real agent to obtain the response and calculate

---

Algorithm 1: Coordinate Descent Algorithm *

Input: System error model M, stop threshold K_{stop} and maximum iterations N.
Output: Optimized parameter $\mathbb{K}^* = \mathbb{K}$
initialization;
$\mathbb{K} \leftarrow$ random 3×1 vector;
$d\mathbb{K} \leftarrow$ positive random 3×1 vector;
count $\leftarrow 0$;
best_error $\leftarrow M(\mathbb{K})$;
while $|d\mathbb{K}|_1 \geq K_{stop}$ **and** count $< N$ **do**
    for $i \leftarrow \{1,2,3\}$ **do**
        $\mathbb{K}[i] \leftarrow \mathbb{K}[i] + d\mathbb{K}[i]$;
        error $\leftarrow M(\mathbb{K})$;
        if error $\leq$ best_error **then**
            $d\mathbb{K}[i] \leftarrow 1.2 * d\mathbb{K}[i]$;
            best_error $\leftarrow$ error;
        else
            $\mathbb{K}[i] \leftarrow \mathbb{K}[i] - 2 * d\mathbb{K}[i]$;
            error $\leftarrow M(\mathbb{K})$;
            if error $\leq$ best_error **then**
                $d\mathbb{K}[i] \leftarrow 1.2 * d\mathbb{K}[i]$;
                best_error $\leftarrow$ error;
            else
                $\mathbb{K}[i] \leftarrow \mathbb{K}[i] + d\mathbb{K}[i]$;
                $d\mathbb{K}[i] \leftarrow 0.8 * d\mathbb{K}[i]$;
            end
        end
    end
    count $\leftarrow$ count $+ 1$;
end

---

the performance metric. Therefore, we used a classic automatic tuning algorithm, called Coordinate Descent Algorithm [85]. It is a metaheuristic algorithm that tries to optimize a concerned control performance metric in the output of the system. The pseudo-code is given in Algorithm 1. The heuristics of the algorithm runs as follows. It starts with a random value for the controller parameter vector $\mathbb{K}$ and a small probing positive step size $d\mathbb{K}$. It then calculates concerned performance metric $M(\mathbb{K})$ of the control system. If the metric is not good enough, then the algorithm probes the neighborhood of $\mathbb{K}$ with a step size of $d\mathbb{K}$. If the metric improves, we update $\mathbb{K}$ and increase the step size $d\mathbb{K}$; otherwise, we try smaller step sizes. We repeat the above process until the step size shrinks below a threshold or until a maximum iteration count is reached.

Note that the Coordinate Descent Algorithm is a metaheuristic algorithm; therefore, it may converge to a local optimum instead of a global optimum, or not converge at all, depending on the initial value. Also, in this chapter, our concerned control performance metric is defined by the following:

$$M(\mathbb{K}) = \frac{1}{n} \sum_{i=1}^{n} \left(|e[i]| + \left| \frac{\Delta e_i}{\Delta t_i} \right| \right), \tag{2.5}$$

where $e[i]$ and $\Delta e_i / \Delta t_i$ are same as defined in (2.2). This definition takes into consideration both; the response oscillation magnitude (as measured by $|e[i]|$) and the oscillation slope (as measured by $|\Delta e_i / \Delta t_i|$), and therefore can reflect the control accuracy and dynamic response quality.

To summarize, our Coordinate Descent Algorithm aims to find the optimal PID control parameter vector

$$M^* = \underset{\mathbb{K}}{\operatorname{argmin}} \, M(\mathbb{K}), \tag{2.6}$$

by minimizing the $M(\mathbb{K})$ of Eq. 2.5, which is a holistic metric of control accuracy and dynamic response quality.

? Why Stepest Descent algorithm is used for parameter tuning?

Several model-based parameter tuning algorithm have been proposed in literature. For example, [82], assumes a second-order linear model of the controlled plant. Similarly, [83] related to the control of Automatic Voltage Regulator (AVR) also assumes that the model of the plant is known apriori. However, for the case of the soft agents, the model of the system is highly nonlinear and theoretically requires an infinite degree of freedom [58] to accurately model the dynamics because of the flexible body. Since it is incredibly challenging to develop an accurate mathematical model of soft agents, therefore we cannot rely on simulations to evaluate the value of performance metric, i.e., tracking error, to find an optimal set of PID parameters. Therefore coordinate descent, a metaheuristic optimization algorithm which requires a fewer number of function evaluations is used. Although one can argue that the model of the soft agents can still be approximated with low-order linear models, but this approach fails to identify the uniqueness of the soft agents as compared to the rigid robots. Therefore, the focus of the proposed work is to study the comparative performance of the model-free control algorithms for the case of soft agents, specifically.

2.4 Experimental Platform, Results and Discussion

In this section, we evaluate the various model-free closed-loop PID-controllers on soft agents via experiments.

2.4.1 Experimental Platform

PneuNet [52] are soft agents capable of producing bending motion on pneumatic actuation. It consists of a series of inflatable chambers mounted on a single base of silicone material and connected through a channel. The silicone base is stiffer as compared to chambers, which causes a bending when chambers are inflate using pressurized air. The fabrication, sensing, and actuation mechanisms are described now.

2.4.1.1 Manufacturing The Experimental Platform

The design of PneuNets molds is publicly available [86]. We used Dragon Skin 10 [87], a silicone material. This material consists of two separate mixtures; Part A and Part B, respectively. The silicone starts curing after both mixtures are combined and placed in the open air for four to eight hours.

We 3D printed the molds and prepared the elastomer by thoroughly mixing Part A and Part B of Dragon Skin 10 in a 1:1 ratio by volume. The printed molds consist of three parts: one part is used to fabricate the bottom of PneuNet, and the other two parts are combined to fabricate the upper chambers. A prepared elastomer is poured into the molds and allowed to be cured in the open air at room temperature for 8 hours. Since the bottom of the actuator needs to be stiffer, a piece of paper is embedded inside it. Once the bottom and upper chambers are cured, they are glued with a liquid elastomer to form a holistic PneuNet. Figure 2.3 shows the whole fabrication process.

2.4.1.2 Pneumatic Actuation and Sensing Mechanisms

We used FlexiForce bending sensors to estimate the bending angle of PneuNets. Such a sensor is a resistor whose resistance changes with the bending angle. We modeled a mapping from resistance values to bending angles using statistical data. The sensor is then attached to the base of our PneuNet, as shown in Fig. 2.4a.

A 12V DC air pump is used as an actuation source for the soft agent. The pump was connected to the PneuNet via an electro-mechanic valve controlled by a MOSFET switch. The valves have a switching period of $T = 1/30$ seconds, and they can be controlled using the PWM signal as input. The pressure at the output of valves is directly proportional to the duty cycle of PWM. Between the bending sensor and the MOSFET switch is the PID-controller. The controller runs on an Arduino Mega 2560 board and outputs PWM signals to adjust the duty cycle of the MOSFET switch. The developed system is shown in Fig. 2.4b.

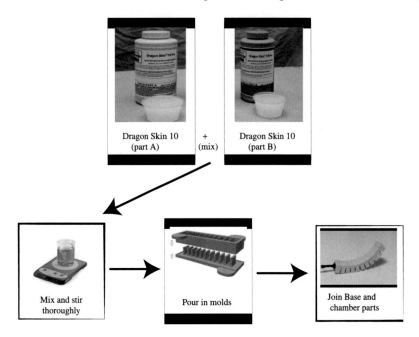

| Dragon Skin 10 (part A) | + (mix) | Dragon Skin 10 (part B) |

| Mix and stir thoroughly | Pour in molds | Join Base and chamber parts |

Fig. 2.3 Flowchart of the process for fabricating the PneuNet used in our experimental platform

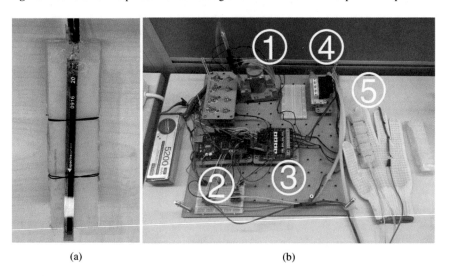

(a)   (b)

Fig. 2.4 **a** Sensing mechanism in experiments to measure the bending angle of the PneuNet (PneuNet: white object, flex sensor: brown strip). **b** Experimental platform: containing **1** an Air pump, **2** a Arduino Mega, **3** MOSFET switches, **4** valves, **5** the bending sensor and the PneuNet

2.4.2 Results and Discussion

We conducted a series of step and sinusoidal input experiments to demonstrate the effectiveness of tuning and control algorithms, respectively. First, the results for the parameter tuning experiments, both manual and automatic, will be presented. The tuned parameters are then used to compare the accuracy of the presented controllers.

2.4.2.1 Parameter Tuning Results

To use the piecewise and fuzzy PID formulations described in Sect. 2.3.1, the entire reference angle range $[0°, 90°]$ is divided into the following three sub-ranges: $[0°, 30°)$, $[30°, 60°)$ and $[60°, 90°]$. The midpoint of each subrange, i.e. $(15°, 45°,$ and $75°)$ is used as the representative point when tuning the control parameter for that subrange. The endpoints of the entire range, i.e., $0°$ and $90°$, are also considered for tuning. Since $0°$ is trivial ($\theta_r = 0°$ is the natural state of the PneuNet), the results are only presented for $\theta_r = 90°$.

In manual control parameter tuning experiments, the PneuNet is given a step input of $\theta_r = 90$ while manually adjusting control parameter vector $\mathbb{K}$. The observed responses are shown in Fig. 2.5. The blue line shows the desired response, whereas the solid red line shows the response that was chosen as the best by the observer because of the short rising time and damped oscillations after the reference angle was reached. In all of the cases, it is observed that setting $K_d \neq 0$ results in oscillations and unstable responses, therefore $K_d = 0$ is an optimal value for the PneuNet actuation. It can be explained in the context of the interpretation of PID-controller terms. The derivative term in PID is essentially a prediction of future behavior of the robotic system. Estimation of the derivative term is trivial for a rigid robotic system since its behavior is mathematically predictable and strictly follow system model. In contrast, soft agents are characterized by a high degree of uncertainty and chaotic motion. Therefore, the estimation of future motion usually results in large variations and errors. These variations manifest themselves in the form of unexpected variation in controller output, thus resulting in erratic behavior. Therefore, a combination of $K_p \neq 0$ and $K_i \neq 0$ is recommended since they can produce the desired performance. This behavior can also be explained in term of the second-order lumped element model of the soft agent. By analyzing a general second-order model of a agentic system driven by a PID controller, the system can become unstable if the natural damping of a soft agent is very small. Since soft agents are inherently flexible, therefore, the internal mechanical damping produced by the elastic material is minimal. Thus using a large value of K_d can render the system unstable. This heuristic was used while manually adjusting the control parameters for $\theta_r = 15°, 45°, 75°$. Table 2.2 summarizes the results of the manual tuning experiments.

In automatic parameter tuning experiments, coordinate descent algorithm described in Sect. 2.3.2 is used to minimize performance metric (2.5). The PneuNet is programmed to sequentially select the reference angle from the set $\{15°, 45°, 75°,$

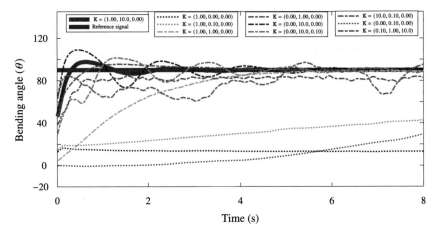

Fig. 2.5 System responses for the manual tuning of control parameters $\mathbb{K}$ for $\theta_r = 90^\circ$. The solid blue line represents the reference response while solid red line represents the response, visually chosen to be the best

Table 2.2 Summary of the parameter tuning experiments. The final tuned values are given

θ_r	Manual tuning			Automatic tuning		
	K_p	K_i	K_d	K_p	K_i	K_d
15	1	1	0	1	2.44	0
45	1	2	0	1.48	3.77	0
75	1	5	0	1.49	2.43	0

90°} for an equal amount of time. Figure 2.6a shows the convergence of control parameters $\mathbb{K}$ in 96 iterations of the coordinate descent. For the described experiment, the set of parameters that produces the minimum value for a performance metric (2.5), is chosen as the best. These results also support the assertion that $K_d = 0$ is an optimal value for the soft actuator.

The convergence of the control parameters for the first subrange $[0^\circ, 30^\circ)$ i.e. $\theta_r = 15^\circ$, using the coordinate descent is shown in Fig. 2.6b. Similar results are shown for the second subrange $[30^\circ, 60^\circ)$ ($\theta_r = 45^\circ$) in Fig. 2.6c and for the third subrange $[60^\circ, 90^\circ)$ ($\theta_r = 75^\circ$) in Fig. 2.6d. These results further strengthen the assertion that $K_d = 0$ is optimal. Table 2.2 gives a summary of the parameter tuning experiments.

2.4.2.2 Comparison Results

The following presents a comparison of the three PID-controllers, presented in Sect. 2.3.1, with manual and automatic tuned control parameters. The controllers are applied with sinusoidally varying reference signals of different frequencies. The

Table 2.3 Comparison between 6 differnet scenarios of PID variants and tuning algorithm is summarized. For each scenario, 25 experiments are performed and summarized as median along with first and third qurantiles of error metric defined in (2.7). In most cases Fuzzy PID with automatic tuning produce best results

Frequency	Manual tuning			Automatic tuning		
	Ordinary PID	Piecewise PID	Fuzzy PID	Ordinary PID	Piecewise PID	Fuzzy PID
0.05	19.03 (16.54, 21.65)	12.29 (10.50, 13.71)	16.37 (13.55, 18.50)	**9.07** (7.49, 12.56)	11.41 (9.29, 12.88)	9.79 (7.86, 12.99)
0.10	18.94 (17.08, 21.51)	17.81 (15.20, 21.13)	19.22 (15.59, 20.99)	17.12 (14.31, 19.50)	13.44 (10.86, 15.32)	**11.05** (9.21, 14.04)
0.20	23.08 (20.25, 24.87)	22.29 (18.10, 25.58)	24.67 (20.40, 26.86)	19.59 (16.77, 21.42)	16.76 (13.24, 20.05)	**15.29** (13.42, 18.48)
0.50	28.40 (24.30, 31.36)	26.49 (21.38, 33.78)	28.48 (26.75, 39.98)	32.74 (31.32, 34.31)	26.15 (22.86, 28.03)	**25.46** (23.01, 29.63)
1.00	42.56 (38.54, 44.64)	36.39 (34.53, 38.14)	35.47 (32.89, 36.35)	39.29 (37.78, 43.43)	34.00 (32.21, 36.40)	**32.32** (31.02, 35.33)
2.00	67.66 (57.90, 70.58)	43.96 (41.53, 48.90)	49.13 (44.89, 60.95)	44.95 (42.14, 46.73)	42.25 (40.60, 45.66)	**42.22** (39.99, 44.58)
5.00	55.78 (53.48, 58.14)	50.01 (39.66, 53.26)	46.20 (41.46, 52.41)	52.78 (44.87, 60.29)	52.03 (46.55, 55.71)	**45.14** (41.80, 39.42)

*Values are written as Median (25, 75th) percentile

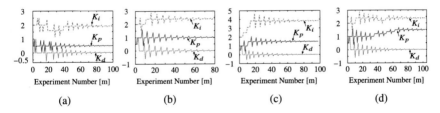

(a)   (b)   (c)   (d)

Fig. 2.6 Control parameter convergance by coordinate descent. Each iteration of coordinate descent represents one expreiment. **a** shows the tuning results for the complete operation range $\theta_r \in [0^o, 90^o]$ i.e., θ_r is sequentially selected from set $\{15^o, 45^o, 75^o, 90^o\}$. The final values are $K_p = 0.5$, $K_i = 1.86$, and $K_d = 0$. **b** shows the results for subrange $\theta_r \in [0^o, 30^o)$ i.e., $\theta_r = 15^o$, the final values are $K_p = 1$, $K_i = 2.44$, and $K_d = 0$. **c** shows the results for subrange $\theta_r = 45^o$, the final values are $K_p = 1.48$, $K_i = 3.77$, and $K_d = 0$. **d** shows the results for subrange $\theta_r \in [60^o, 90^o)$ i.e., $\theta_r = 75^o$, the final values are $K_p = 1.49$, $K_i = 2.43$, and $K_d = 0$

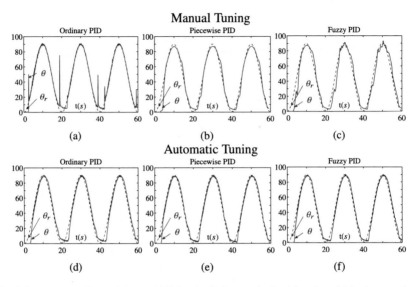

Fig. 2.7 Response of one trial (out of 25) for six distinct methods with a sinusoidal reference signal of 0.05 Hz. At low frequencies, the controller can accurately track time varying reference signals, hence the low value of the error metric (2.7)

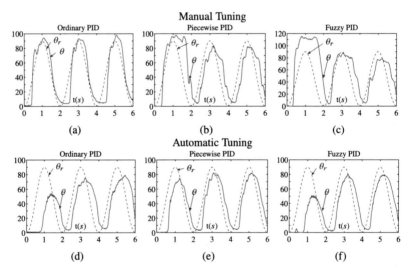

Fig. 2.8 Response of one trial (out of 25) for six distinct methods with a sinusoidal reference signal of 0.5 Hz. At increase in frequency of above 0.05 Hz the output begins to lag the input due to the slow reposne of the PneuNet. The error metric of (2.7) is high as compared to the 0.05 Hz

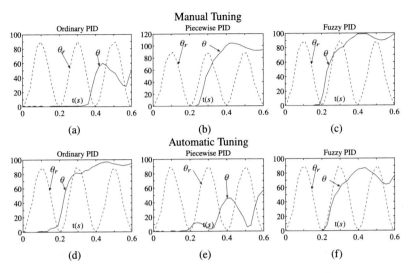

Fig. 2.9 Response of one trial (out of 25) for six distinct methods with a sinusoidal reference signal of 5 Hz. At such a high frequency, the physical property (e.g. inertia) of soft muscles constraint the frequency at which it can track input sinusoidal reference. The soft muscle is completely unable to track the reference signal

controllers are compared based on their ability to accurately track time-varying reference signals. The tracking accuracy of a controller is calculated using the following error metric:

$$E = \frac{1}{n} \sum_{i=1}^{n} \left| \theta_r[i] - \theta[i] \right|, \tag{2.7}$$

where $\theta_r[i]$ is a sinusoidally varying reference angle and $\theta[i]$ is the PneuNet bending angle from the sensor.

Comparisons between the following six distinct PID controllers: {Ordinary, Piecewise, Fuzzy} PID × {Manual, Automatic} is discussed now. The tuned control parameters summarized in Table 2.2 are used to perform all of the experiments. Each method was tested using seven reference signal frequencies of progressively increasing distinct values, varying from 0.05 Hz to 5 Hz. A total of 25 trials were performed for each experimental case. The response of one trial (out of 25) on tracking the signal frequency of 0.05 Hz is shown in Fig. 2.7. Due to the low frequency of the reference signal, the PneuNet can accurately track the time-varying reference angle. Statistics on the error distributions in the 25 trials for each experiment are shown in Fig. 2.10. It can be seen that in most cases, the automatically tuned parameters usually provided better results than the manually tuned parameters. A similar trend can be observed for the Piecewise and Fuzzy PID-controllers, i.e., the error metric is of a smaller value as compared to the Ordinary PID. These obser-

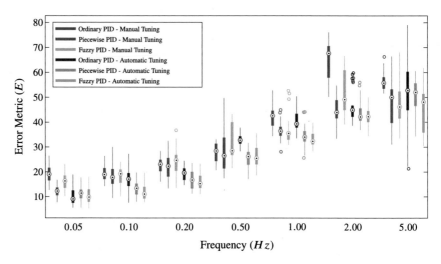

Fig. 2.10 Summary of the tracking error comparision of six distinct experiments at frequencies ranging from 0.05 Hz to 5 Hz. The dots shows the median value while box edges shows 25 and 75th percentile. At higher frequencies, the error metric defined by (2.7) increase

vations are consistent with the theory of PID-controllers. Since Piecewise and Fuzzy PID-controllers continually adapt the parameters according to the reference angle, they provide superior performance as compared to the ordinary PID-controller. At very high frequency, the performance deteriorates significantly; therefore, in that fuzzy might exhibit high error as compared to other methods. Similarly, at low frequency, the system can easily track reference signal; therefore, all methods produce a similar performance.

Then moving toward higher frequency in Fig. 2.10 and Table 2.3, it can be observed that the error metric began to increase gradually. It can be attributed to the quickly varying reference angle and the slow response rate of the PneuNet. The mechanical properties of the PneuNet (i.e., inertia and the response rate of hydraulic systems) contribute to the slow response, in the case of a sudden change in reference angle. From the experiments, it was observed that the PneuNet could track reference signals of up to 0.5 Hz. Higher frequencies will cause error metrics of large value. The response in one trial (out of 25) for a reference signal frequency of 0.5 Hz is shown in Fig. 2.8. In this case, the PneuNet is barely able to track the reference signal, with a large lag between output and input. Similar results for 5 Hz frequency are shown in Fig. 2.9.

2.5 Conclusion

In this chapter, an experimental comparison between three variants of PID-controllers for a soft agent has been presented. Since PID-controllers require tuning of their parameter to achieve desired dynamic response, we presented a comprehensive overview of manual and automatic parameter tuning algorithms. We showed that an automatic parameter tuning algorithm, such as a coordinate descent, could be used to optimize a performance metric, e.g., rise-time, overshoot, and settling time. A comparison was made between six variants is presented. It was empirically demonstrated that automatically tuned parameters could generally produce better results than manually tuned parameters. Furthermore, the Piecewise and Fuzzy PID were shown to be better suited for soft agents for most cases because of their flexibility in continually adapting the control parameters. Before ending this section as well as this chapter, it is worth mentioning that this is the first systematical evaluation of various PID-controller variants use extensive data from real experiments and the first time that identifies a critical difference between rigid-body and soft agents.

References

1. X. Jiang and S. Li, "Beetle antennae search without parameter tuning (bas-wpt) for multi-objective optimization," *arXiv preprint* arXiv:1711.02395, 2017.
2. D. Chen, S. Li, F.-J. Lin, and Q. Wu, "New super-twisting zeroing neural-dynamics model for tracking control of parallel robots: A finite-time and robust solution," *IEEE transactions on cybernetics*, 2019.
3. D. Chen, S. Li, W. Li, and Q. Wu, "A multi-level simultaneous minimization scheme applied to jerk-bounded redundant robot manipulators," *IEEE Transactions on Automation Science and Engineering*, 2019.
4. D. Chen, Y. Zhang, and S. Li, "Tracking control of robot manipulators with unknown models: A jacobian-matrix-adaption method," *IEEE Transactions on Industrial Informatics*, vol. 14, no. 7, pp. 3044–3053, 2017.
5. V. Vikas, P. Grover, and B. Trimmer, "Model-free control framework for multi-limb soft robots," in *Intelligent Robots and Systems (IROS), 2015 IEEE/RSJ International Conference on*, pp. 1111–1116, IEEE, 2015.
6. H.-T. Lin, G. G. Leisk, and B. Trimmer, "Goqbot: a caterpillar-inspired soft-bodied rolling robot," *Bioinspiration & biomimetics*, vol. 6, no. 2, p. 026007, 2011.
7. M. Calisti, M. Giorelli, G. Levy, B. Mazzolai, B. Hochner, C. Laschi, and P. Dario, "An octopus-bioinspired solution to movement and manipulation for soft robots," *Bioinspiration & biomimetics*, vol. 6, no. 3, p. 036002, 2011.
8. A. Marchese, C. Onal, and D. Rus, "Autonomous soft robotic fish capable of escape maneuvers using fluidic elastomer actuators," *Soft Robotics*, vol. 1, no. 1, pp. 75–87, 2014.
9. P. Polygerinos, S. Lyne, Z. Wang, L. Nicolini, B. Mosadegh, G. Whitesides, and C. Walsh, "Towards a soft pneumatic glove for hand rehabilitation," in *Intelligent Robots and Systems (IROS), 2013 IEEE/RSJ International Conference on*, pp. 1512–1517, IEEE, 2013.
10. F. Largilliere, V. Verona, E. Coevoet, M. Sanz-Lopez, J. Dequidt, and C. Duriez, "Real-time control of soft-robots using asynchronous finite element modeling," in *Robotics and Automation (ICRA), 2015 IEEE International Conference on*, pp. 2550–2555, IEEE, 2015.
11. F. Faure, C. Duriez, H. Delingette, J. Allard, B. Gilles, S. Marchesseau, H. Talbot, H. Courte-cuisse, G. Bousquet, I. Peterlik, *et al.*, "Sofa: A multi-model framework for interactive physical simulation," in *Soft Tissue Biomechanical Modeling for Computer Assisted Surgery*, pp. 283–321, Springer, 2012.

12. C. Duriez, "Control of elastic soft robots based on real-time finite element method," in *Robotics and Automation (ICRA), 2013 IEEE International Conference on*, pp. 3982–3987, IEEE, 2013.

13. A. Marchese, R. Tedrake, and D. Rus, "Dynamics and trajectory optimization for a soft spatial fluidic elastomer manipulator," *The International Journal of Robotics Research*, vol. 35, no. 8, pp. 1000–1019, 2016.

14. F. Renda, M. Giorelli, M. Calisti, M. Cianchetti, and C. Laschi, "Dynamic model of a multi-bending soft robot arm driven by cables," *IEEE Transactions on Robotics*, vol. 30, no. 5, pp. 1109–1122, 2014.

15. I. Gravagne, C. Rahn, and I. Walker, "Large deflection dynamics and control for planar continuum robots," *IEEE/ASME transactions on mechatronics*, vol. 8, no. 2, pp. 299–307, 2003.

16. A. H. Khan, S. Li, X. Zhou, Y. Li, M. U. Khan, X. Luo, and H. Wang, "Neural & bio-inspired processing and robot control," *Frontiers in neurorobotics*, vol. 12, 2018.

17. X. Jiang, S. Li, B. Luo, and Q. Meng, "Source exploration for an under-actuated system: A control-theoretic paradigm," *IEEE Transactions on Control Systems Technology*, 2019.

18. Y. Zhang, S. Li, and X. Jiang, "Near-optimal control without solving hjb equations and its applications," *IEEE Transactions on Industrial Electronics*, vol. 65, no. 9, pp. 7173–7184, 2018.

19. X. Jiang and S. Li, "Plume front tracking in unknown environments by estimation and control," *IEEE Transactions on Industrial Informatics*, vol. 15, no. 2, pp. 911–921, 2018.

20. A. T. Khan and S. Li, "A survey on blockchain technology and its potential applications in distributed control and cooperative robots," *arXiv preprint* arXiv:1812.05452, 2018.

21. A. H. Khan, S. Li, and X. Bin, "Bas-swarm: A nature-inspired metaheuristic algorithm with applications in machine learning," *Soft Computing*, vol. 1, no. 1, p. 1, 2019.

22. A. H. Khan, X. Cao, S. Li, and C. Luo, "Using social behavior of beetles to establish a computational model for operational management," *IEEE Transactions on Computational Social Systems*, vol. 7, no. 2, pp. 492–502, 2020.

23. A. H. Khan and S. Li, "Tracking control of redundant manipulator under active remote center of motion constraints: An rnn-based metaheuristic approach," *SCIENCE CHINA Information Sciences*, 2019.

24. A. H. Khan, S. Li, D. Chen, and L. Liao, "Tracking control of redundant mobile manipulator: An rnn based metaheuristic approach," *Neurocomputing*, 2020.

25. Q. Wu, X. Shen, Y. Jin, Z. Chen, S. Li, A. H. Khan, and D. Chen, "Intelligent beetle antenna search for uav sensing and avoidance of obstacles," *Sensors*, vol. 19, no. 8, p. 1758, 2019.

26. A. H. Khan, X. Cao, S. Li, V. N. Katsikis, and L. Liao, "Bas-adam: An adam based approach to improve the performance of beetle antennae search optimizer," *IEEE/CAA Journal of Automatica Sinica*, vol. 7, no. 2, pp. 461–471, 2020.

27. F. Ni, A. Henning, K. Tang, and L. Cai, "Soft damper for quick stabilization of soft robotic actuator," in *Real-time Computing and Robotics (RCAR), IEEE International Conference on*, pp. 466–471, IEEE, 2016.

28. Y. Wei, Y. Chen, T. Ren, Q. Chen, C. Yan, Y. Yang, and Y. Li, "A novel, variable stiffness robotic gripper based on integrated soft actuating and particle jamming," *Soft Robotics*, vol. 3, no. 3, pp. 134–143, 2016.

29. Y. Li, Y. Chen, T. Ren, and Y. Hu, "Passive and active particle damping in soft robotic actuators," in *Robotics and Automation (ICRA), 2018 IEEE International Conference on*, pp. 1547–1552, IEEE, 2018.

30. M. Luo, E. H. Skorina, W. Tao, F. Chen, S. Ozel, Y. Sun, and C. D. Onal, "Toward modular soft robotics: Proprioceptive curvature sensing and sliding-mode control of soft bidirectional bending modules," *Soft robotics*, vol. 4, no. 2, pp. 117–125, 2017.

31. S. Terryn, J. Brancart, D. Lefeber, G. Van Assche, and B. Vanderborght, "Self-healing soft pneumatic robots," *Sci. Robot.*, vol. 2, p. eaan4268, 2017.

32. G. Gerboni, A. Diodato, G. Ciuti, M. Cianchetti, and A. Menciassi, "Feedback control of soft robot actuators via commercial flex bend sensors," *IEEE/ASME Transactions on Mechatronics*, 2017.

33. A. H. Khan, Z. Shao, S. Li, Q. Wang, and N. Guan, "Which is the best pid variant for pneumatic soft robots? an experimental study," *IEEE/CAA Journal of Automatica Sinica*, vol. 6, no. 1, p. 1, 2019.

34. K. Ang, G. Chong, and Y. Li, "Pid control system analysis, design, and technology," *IEEE transactions on Control Systems Technology*, vol. 13, no. 4, pp. 559–576, 2005.

35. K. J. Åström and T. Hägglund, *PID controllers: theory, design, and tuning*, vol. 2. Instrument society of America Research Triangle Park, NC, 1995.

36. K.-S. Tang, K. F. Man, G. Chen, and S. Kwong, "An optimal fuzzy pid controller," *IEEE Transactions on Industrial Electronics*, vol. 48, no. 4, pp. 757–765, 2001.

37. X. Luo, M. Zhou, S. Li, Y. Xia, Z.-H. You, Q. Zhu, and H. Leung, "Incorporation of efficient second-order solvers into latent factor models for accurate prediction of missing qos data," *IEEE transactions on cybernetics*, vol. 48, no. 4, pp. 1216–1228, 2017.

38. X. Luo, M. Zhou, S. Li, and M. Shang, "An inherently nonnegative latent factor model for high-dimensional and sparse matrices from industrial applications," *IEEE Transactions on Industrial Informatics*, vol. 14, no. 5, pp. 2011–2022, 2017.

39. S. Li, Z. Wang, and Y. Li, "Using laplacian eigenmap as heuristic information to solve nonlinear constraints defined on a graph and its application in distributed range-free localization of wireless sensor networks," *Neural processing letters*, vol. 37, no. 3, pp. 411–424, 2013.

40. S. Li, R. Kong, and Y. Guo, "Cooperative distributed source seeking by multiple robots: Algorithms and experiments," *IEEE/ASME Transactions on mechatronics*, vol. 19, no. 6, pp. 1810–1820, 2014.

41. A. T. Khan, S. L. Senior, P. S. Stanimirovic, and Y. Zhang, "Model-free optimization using eagle perching optimizer," *arXiv preprint* arXiv:1807.02754, 2018.

42. A. H. Khan, S. Li, and X. Luo, "Obstacle avoidance and tracking control of redundant robotic manipulator: An rnn based metaheuristic approach," *IEEE Transactions on Industrial Informatics*, 2019.

43. Y. Zhang, S. Li, J. Zou, and A. H. Khan, "A passivity-based approach for kinematic control of redundant manipulators with constraints," *IEEE Trans. on Ind. Informatics*, 2019.

44. L. Xiao, S. Li, F.-J. Lin, Z. Tan, and A. H. Khan, "Zeroing neural dynamics for control design: comprehensive analysis on stability, robustness, and convergence speed," *IEEE Transactions on Industrial Informatics*, vol. 15, no. 5, pp. 2605–2616, 2018.

45. L. Jin and S. Li, "Distributed task allocation of multiple robots: A control perspective," *IEEE Transactions on Systems, Man, and Cybernetics: Systems*, vol. 48, no. 5, pp. 693–701, 2016.

46. L. Jin, S. Li, H. M. La, and X. Luo, "Manipulability optimization of redundant manipulators using dynamic neural networks," *IEEE Transactions on Industrial Electronics*, vol. 64, no. 6, pp. 4710–4720, 2017.

47. S. Li, Y. Guo, and B. Bingham, "Multi-robot cooperative control for monitoring and tracking dynamic plumes," in *2014 IEEE International Conference on Robotics and Automation (ICRA)*, pp. 67–73, IEEE, 2014.

48. S. Li and Y. Guo, "Distributed source seeking by cooperative robots: All-to-all and limited communications," in *2012 IEEE International Conference on Robotics and Automation*, pp. 1107–1112, IEEE, 2012.

49. S. Li, Y. Lou, and B. Liu, "Bluetooth aided mobile phone localization: a nonlinear neural circuit approach," *ACM Transactions on Embedded Computing Systems (TECS)*, vol. 13, no. 4, p. 78, 2014.

50. B. Tondu, "Modelling of the mckibben artificial muscle: A review," *Journal of Intelligent Material Systems and Structures*, vol. 23, pp. 225–253, 2012.

51. M. Doumit, A. Fahim, and M. Munro, "Analytical modeling and experimental validation of the braided pneumatic muscle," *IEEE transactions on robotics*, vol. 25, no. 6, pp. 1282–1291, 2009.

52. B. Mosadegh, P. Polygerinos, C. Keplinger, S. Wennstedt, R. Shepherd, U. Gupta, J. Shim, K. Bertoldi, C. Walsh, and G. Whitesides, "Pneumatic networks for soft robotics that actuate rapidly," *Advanced Functional Materials*, 2013.

53. K. Galloway, P. Polygerinos, C. Walsh, and R. Wood, "Mechanically programmable bend radius for fiber-reinforced soft actuators," in *Advanced Robotics (ICAR), 2013 16th International Conference on*, pp. 1–6, IEEE, 2013.

54. Y. Yang, Y. Chen, Y. Li, M. Z. Chen, and Y. Wei, "Bioinspired robotic fingers based on pneumatic actuator and 3d printing of smart material," *Soft robotics*, vol. 4, no. 2, pp. 147–162, 2017.

55. X. Luo, J. Sun, Z. Wang, S. Li, and M. Shang, "Symmetric and nonnegative latent factor models for undirected, high-dimensional, and sparse networks in industrial applications," *IEEE Transactions on Industrial Informatics*, vol. 13, no. 6, pp. 3098–3107, 2017.

56. X. Luo, M. Zhou, S. Li, Z. You, Y. Xia, and Q. Zhu, "A nonnegative latent factor model for large-scale sparse matrices in recommender systems via alternating direction method," *IEEE transactions on neural networks and learning systems*, vol. 27, no. 3, pp. 579–592, 2015.

57. R. Martinez, J. Branch, C. Fish, L. Jin, R. Shepherd, R. Nunes, Z. Suo, and G. Whitesides, "Robotic tentacles with three-dimensional mobility based on flexible elastomers," *Advanced Materials*, vol. 25, no. 2, pp. 205–212, 2013.

58. C. Della Santina, R. K. Katzschmann, A. Bicchi, and D. Rus, "Dynamic control of soft robots interacting with the environment," 2018.

59. E. H. Skorina, M. Luo, W. Tao, F. Chen, J. Fu, and C. D. Onal, "Adapting to flexibility: Model reference adaptive control of soft bending actuators," *IEEE Robotics and Automation Letters*, vol. 2, no. 2, pp. 964–970, 2017.

60. C. Keplinger, T. Li, R. Baumgartner, Z. Suo, and S. Bauer, "Harnessing snap-through instability in soft dielectrics to achieve giant voltage-triggered deformation," *Soft Matter*, vol. 8, no. 2, pp. 285–288, 2012.

61. I. A. Anderson, T. A. Gisby, T. G. McKay, B. M. O'Brien, and E. P. Calius, "Multi-functional dielectric elastomer artificial muscles for soft and smart machines," *Journal of Applied Physics*, vol. 112, no. 4, p. 041101, 2012.

62. J. Overvelde, T. Kloek, J. D'haen, and K. Bertoldi, "Amplifying the response of soft actuators by harnessing snap-through instabilities," *Proceedings of the National Academy of Sciences*, vol. 112, no. 35, pp. 10863–10868, 2015.

63. W. Felt, K. Chin, and C. Remy, "Contraction sensing with smart braid mckibben muscles," *IEEE/ASME Transactions on Mechatronics*, vol. 21, no. 3, pp. 1201–1209, 2016.

64. Y.-L. Park, B.-R. Chen, C. Majidi, R. Wood, R. Nagpal, and E. Goldfield, "Active modular elastomer sleeve for soft wearable assistance robots," in *Intelligent Robots and Systems (IROS), 2012 IEEE/RSJ International Conference on*, pp. 1595–1602, IEEE, 2012.

65. Y.-L. Park, C. Majidi, R. Kramer, P. Bérard, and R. Wood, "Hyperelastic pressure sensing with a liquid-embedded elastomer," *Journal of Micromechanics and Microengineering*, vol. 20, no. 12, p. 125029, 2010.

66. A. Veale, I. Anderson, and S. Xie, "The smart peano fluidic muscle: a low profile flexible orthosis actuator that feels pain," in *SPIE Smart Structures and Materials*, pp. 94351V–94351V, International Society for Optics and Photonics, 2015.

67. H. Lin, F. Guo, F. Wang, and Y.-B. Jia, "Picking up a soft 3d object by "feeling" the grip," *The International Journal of Robotics Research*, vol. 34, no. 11, pp. 1361–1384, 2015.

68. X. Luo, H. Wu, H. Yuan, and M. Zhou, "Temporal pattern-aware qos prediction via biased non-negative latent factorization of tensors," *IEEE transactions on cybernetics*, 2019.

69. I. Galiana, F. Hammond, R. Howe, and M. Popovic, "Wearable soft robotic device for post-stroke shoulder rehabilitation: Identifying misalignments," in *Intelligent Robots and Systems (IROS), 2012 IEEE/RSJ International Conference on*, pp. 317–322, IEEE, 2012.

70. M. Zhu, W. Xu, and L. K. Cheng, "Esophageal peristaltic control of a soft-bodied swallowing robot by the central pattern generator," *IEEE/ASME Transactions on Mechatronics*, vol. 22, no. 1, pp. 91–98, 2017.

71. H. In, U. Jeong, H. Lee, and K.-J. Cho, "A novel slack-enabling tendon drive that improves efficiency, size, and safety in soft wearable robots," *IEEE/ASME Transactions on Mechatronics*, vol. 22, no. 1, pp. 59–70, 2017.

72. X. Luo, M. Zhou, Y. Xia, Q. Zhu, A. C. Ammari, and A. Alabdulwahab, "Generating highly accurate predictions for missing qos data via aggregating nonnegative latent factor models," *IEEE transactions on neural networks and learning systems*, vol. 27, no. 3, pp. 524–537, 2015.

73. X. Luo, M. Zhou, Y. Xia, and Q. Zhu, "An efficient non-negative matrix-factorization-based approach to collaborative filtering for recommender systems," *IEEE Transactions on Industrial Informatics*, vol. 10, no. 2, pp. 1273–1284, 2014.

74. Y. Hao, Z. Gong, Z. Xie, S. Guan, X. Yang, Z. Ren, T. Wang, and L. Wen, "Universal soft pneumatic robotic gripper with variable effective length," in *Control Conference (CCC), 2016 35th Chinese*, pp. 6109–6114, IEEE, 2016.

75. S. Li, J. He, Y. Li, and M. U. Rafique, "Distributed recurrent neural networks for cooperative control of manipulators: A game-theoretic perspective," *IEEE transactions on neural networks and learning systems*, vol. 28, no. 2, pp. 415–426, 2016.

76. S. Li, S. Chen, and B. Liu, "Accelerating a recurrent neural network to finite-time convergence for solving time-varying sylvester equation by using a sign-bi-power activation function," *Neural processing letters*, vol. 37, no. 2, pp. 189–205, 2013.

77. S. Li and Y. Li, "Nonlinearly activated neural network for solving time-varying complex sylvester equation," *IEEE Transactions on Cybernetics*, vol. 44, no. 8, pp. 1397–1407, 2013.

78. S. Li, Y. Zhang, and L. Jin, "Kinematic control of redundant manipulators using neural networks," *IEEE transactions on neural networks and learning systems*, vol. 28, no. 10, pp. 2243–2254, 2016.

79. S. Li, Z.-H. You, H. Guo, X. Luo, and Z.-Q. Zhao, "Inverse-free extreme learning machine with optimal information updating," *IEEE transactions on cybernetics*, vol. 46, no. 5, pp. 1229–1241, 2015.

80. S. Li, B. Liu, and Y. Li, "Selective positive–negative feedback produces the winner-take-all competition in recurrent neural networks," *IEEE transactions on neural networks and learning systems*, vol. 24, no. 2, pp. 301–309, 2012.

81. M. Loepfe, C. Schumacher, U. Lustenberger, and W. Stark, "An untethered, jumping roly-poly soft robot driven by combustion," *Soft Robotics*, vol. 2, no. 1, pp. 33–41, 2015.

82. Z. Bingul and O. Karahan, "Comparison of pid and fopid controllers tuned by pso and abc algorithms for unstable and integrating systems with time delay," *Optimal Control Applications and Methods*, vol. 39, no. 4, pp. 1431–1450, 2018.

83. Z. Bingul and O. Karahan, "A novel performance criterion approach to optimum design of pid controller using cuckoo search algorithm for avr system," *Journal of the Franklin Institute*, vol. 355, no. 13, pp. 5534–5559, 2018.

84. J. T. Agee, Z. Bingul, and S. Kizir, "Tip trajectory control of a flexible-link manipulator using an intelligent proportional integral (ipi) controller," *Transactions of the Institute of Measurement and Control*, vol. 36, no. 5, pp. 673–682, 2014.

85. S. J. Wright, "Coordinate descent algorithms," *Mathematical Programming*, vol. 151, no. 1, pp. 3–34, 2015.

86. D. Holland, E. Park, P. Polygerinos, G. Bennett, and C. Walsh, "The soft robotics toolkit: Shared resources for research and design," *Soft Robotics*, vol. 1, no. 3, pp. 224–230, 2014.

87. "Smooth-on inc." https://www.smooth-on.com/tb/files/DRAGON_SKIN_SERIES_TB.pdf. Accessed: 2018-08-28.

Chapter 3
A Novel Damping Mechanism for Soft Agents with Structural Uncertainty

Abstract Chapter two of this brief discussed the techniques to regulate the dynamic response of soft agents using a model-free PID controller. This chapter addresses another critical issue posed by soft agents. Fabricated using silicones, soft agents, are highly elastic systems with the advantage of inherent flexibility, compliance, and safety in human interaction. However, because of their flexible bodies, they oscillate vigorously, when deactuated, before settling down. These oscillations might compromise the structural integrity of a soft agent with time. So far, there is a very little investigation on the passive and active oscillation damping methods for the soft agents. In this work, we present the design of a 6-chambered parallel soft agent and propose an effective active damping method by a smart distribution of the 6 actuation chambers. Experimental verification of the effectiveness of the proposed damping method is conducted on the proposed parallel soft agent. It is shown that the proposed method provides a high degree of oscillation damping thus prolonging the actuator life. Since the proposed method uses the components of the soft agent itself to actively create oscillation damping, there is no additional mechanical overhead.

3.1 Introduction

Soft agents have attracted a great research attention in recent years and have demonstrated their potential application in practical systems [1, 2, 2–5]. Soft agents are used as end-effector of the traditioal rigid manipulators [6–14]. Therefore, the practical applications of soft agents require them to be fast and steady [15–21]. High speed increases productivity while steadiness reduces undesirable effects e.g. overshooting, vibrations. For traditional rigid robots, such undesirable effects are naturally reduced due to use of stiff materials and rigorously studied control methods to provide active damping [22]. On the other hand, owing to their flexible bodies, soft agent usually exhibit large oscillations on deactuation or a sudden change in the control signal. Soft agents are inspired by soft-bodied animals existing in nature e.g. worms, octopus etc [23]. They can easily locomote and interact with the irregular environment due to their flexible and compliant structure, without causing any damage to the

© The Author(s), under exclusive license to Springer Nature Singapore Pte Ltd. 2021
A. H. Khan et al., *Management and Intelligent Decision-Making in Complex Systems: An Optimization-Driven Approach*, https://doi.org/10.1007/978-981-15-9392-5_3

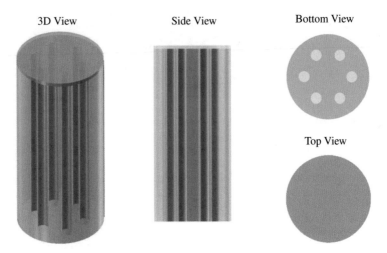

Fig. 3.1 3D model of the 6-chambered parallel soft agent. Left: 3D view of the model, top right: side view, and bottom right: bottom view of the soft agent model

environment [24]. Whereas, traditional rigid robots require complex sensing mechanisms and advanced control theory to safely interact with delicate objects. Soft agents offer great promises in simplifying the problem of safe human-robot interaction [25–34]. But the problem posed by the oscillations need to be addressed.

Soft Pneumatic Actuators (SPAs) [35, 36] are the most common type of soft actuator which have been widely studied and applied in industrial and rehabilitation applications [1, 2, 37–39]. These actuators use pneumatics for actuation and gained popularity because of their fast response rate, simple design, ease of fabrication, and low cost. SPAs consist of several inflatable chambers and actuation is produced by inserting high-pressure air into these chambers. High-pressure increase chamber volume producing motion in the SPA. Several different designs of SPAs are proposed in literature i.e. linear actuators [40], bending actuators [36, 41]. In this chapter, we consider a 6-chambered parallel soft agent as shown in Fig. 3.1.

Despite the advantages offered by soft agents as compared to the traditional rigid robots, they also pose several challenges. Most important among those challenges is accurate and robust control of the motion of soft agents. Since soft agents are made entirely of soft materials, their flexible structure undergo large overshoot and oscillations when suddenly deactuated at high-pressure air. These oscillations happen because, in the absence of any external damping, the natural stiffness of the flexible material is very small. Figure 3.2 shows the oscillations amplitude caused deactuation. The oscillations have an amplitude of about 20° and settling time of about 0.9 seconds. Such large oscillations in the soft agent will not only reduce their viability and efficiency in industrial applications but can also cause undesirable effects such as

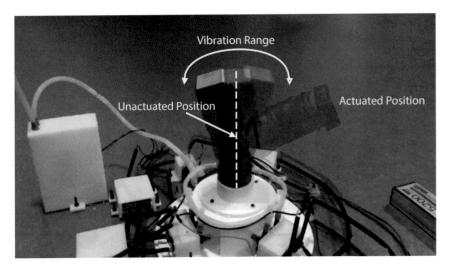

Fig. 3.2 The oscillations produced in soft agent on deactuation

- Increase in operating time.
- Damage to the delicate objects present in surrounding of soft agent.
- Cause wear and tear of soft agent, reducing the lifetime.

These characteristics will greatly impact their usefulness in time-critical industrial applications, where accuracy and robustness are of utmost importance.

Mechanical system dissipates their kinetic and potential energy, because of damping, in the form of heat when deactuated. Every material has internal damping depending on its stiffness e.g. rigid materials have high damping as compared to soft material [42]. For soft materials, the internal damping is not enough to suppress oscillations. An external damping method needs to be used to achieve the desired level of oscillation damping and quick steady-state stabilization. External damping is further classified into passive and active damping [42–47]. Ni et al. [48, 49] propose a passive damping technique by attaching an additional mechanical damper along with the soft agent. The proposed method is able to provide the desired level of damping but the use of additional components make the system bulky. Active damping method includes the active use of actuation signal to create damping effect. Li et al. [50] proposes the use of a seperate partical chamber attached to the soft agent. The damping is created by applying a negative vaccum suction pressure to the particle chamber. The particle chamber augment the energy discsipation by creating frictional and collision forces.

In this work, we propose an active damping approach by smartly distributing the inflatable chambers in the body of the soft agent. The design of the proposed 6-chambered parallel soft agent shown in Fig. 3.1. The soft agent consists of 6 linear empty chambers symmetrically distributed in the circular pattern inside the soft agent's body. When a chamber is inflated, the volume of that chamber increases, forcing the soft agent to bend in the opposite direction. Note that, in absence of any active damping, the chamber radially opposite to the actuated chamber will always be deactuated. If both radially opposite chambers are actuated together, they will cancel each others bending effect. We leverage this cancellation property of radially opposite chambers to create active oscillation damping during actuation and deactuation.

The rest of the chapter is distributed as follow: Section 3.2 describes the design, fabrication, actuation and sensing mechanism of the 6-chambered parallel soft agent, Sect. 3.3 describes the experimental platform and evaluation methodology, Sect. 3.4 presents the experimental results with Sect. 3.5 concluding the chapter.

3.2   Soft Agent Design and Damping Mechanism

In this section, the design, fabrication, sensing and actuation mechanism of the 6-chambered parallel soft agent.

3.2.1   Actuator Design

The previous works [48–50] on oscillation damping of soft agents add additional mechanical components, to create active or passive damping. In this work, we propose a novel design of the soft actuator such that the different components of the soft agents are capable of generating damping for each other, without any additional mechanical overhead. The 3D design of the proposed 6-chambered parallel soft agent as shown in Fig 3.1. The parallel soft agent has a cylindrical soft body, embedding six parallel linear chambers. The linear chambers are distributed evenly in a circular pattern inside the body of the soft agent. The key to the active damping lies in distributing the chambers in a circular pattern so that radially opposite chambers will provide oscillation damping by motion cancellation effect as explained later in this section. One end of the cylindrical soft agent is fixed to a solid base. In deactuated state, the soft agent remains vertical. When one of the chambers is actuated, the volume of that chamber increases and the soft agent bend in the opposite direction.

The soft agent was fabricated using Dragon Skin 30 [51] silicone. The length and outer radius of the soft agent are 10cm and 4cm respectively, whereas the radius of each inner chamber is 5mm. We designed the molds as shown in Fig. 3.3. The liquid silicone was poured in 3D printed molds and allowed to be cured in the open air for about 8 hours. After curing of silicone was complete, the solidified soft agent was removed from the molds. The agent is wrapped in fabric to prevent the damage to the soft agent on a sudden application of high air pressure.

Fig. 3.3 3D drawing of the molds used to cast the 6-chambered parallel soft agent. Left: base of the mold, top right: cap of the mold, and bottom right: the wall to be inserted in mold base

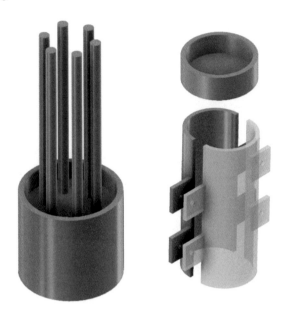

3.2.2 Soft Agent's Model

The soft actuator design presented in the last section consists of 6 chambers. The chambers are evenly distirbuted along the circumference of the soft agent in a circular pattern. When a single channel is actuated, the soft agent produce a simple bending motion. This bending motion of the soft agent can be expressed in term of the bending angle θ of its top surface relative to its initial horizontal position. The bending angle θ is related to the air pressure P inside the air chamber. The relation between P and θ can be drived using the Lagrangian $\mathbb{L}$ of the soft agent. The Lagrangian of a system is defined as

$$\mathbb{L} = \mathbb{T} - \mathbb{V}, \tag{3.1}$$

where $\mathbb{T}$ is the total kinetic energy and $\mathbb{V}$ is the potential energy present in the system. Using the Lagrangian $\mathbb{L}$ the dynamic equation of the system can be defined as

$$\frac{d}{dt}\frac{\partial \mathbb{L}}{\partial \dot{\theta}} - \frac{\partial \mathbb{L}}{\partial \theta} = \tau - b\dot{\theta}, \tag{3.2}$$

where τ is the generalized input force and b models the frictional forces prenet in the system.

The total potential energy $\mathbb{V}$ of the system is mainly contributed by the elastic potential energy. The elastic potential energy of a deformable system is given by the following relation

$$\mathbb{V} = \frac{1}{2}V E \epsilon^2, \tag{3.3}$$

where E is the Young's modulus, V is the volume of the material undergoing deformation and ϵ is the strain present in the soft agent as the result of bending motion. As shown in [52], the relation between strain ϵ and bending angle θ can be approximated to be linear for soft bending agent i.e. $\epsilon = k\theta$, where k is a constant of proportionality. Therefore the total potential energy of the system can be expressed as

$$\mathbb{V} = \frac{1}{2} V E k^2 \theta^2.$$

The generalized input force τ is given by $\partial \mathbb{V}/\partial \theta$, using relation (3.3)

$$\tau = V E k^2 \epsilon.$$

Since the actual physical input to the system is air pressure P, we are interested in the relation of generalized force τ in term of P. As shown by [53] the strain produced in a bending soft agent is directly proportional to its internal air pressure P i.e. $\epsilon = cP$. Here c is a constant of proportionality between strain ϵ and air pressure P. Therefore

$$\tau = V E c P.$$

Now we will calculate the total kinetic energy $\mathbb{T}$ present in the system. The total kinetic energy is mainly contributed by the rotational kinetic energy of the soft agent

$$T = \frac{1}{2} I \dot{\theta}^2,$$

where I is the rotational inertia of the soft agent. Replacing the derived values of $\mathbb{T}$ and $\mathbb{V}$ in (3.1), the lagrangian becomes

$$L = \frac{1}{2} I \dot{\theta}^2 - \frac{1}{2} V E k^2 \theta^2.$$

Putting the values of $\mathbb{L}$ and τ in (3.2), we get

$$I \ddot{\theta} + V E k^2 \theta = V E c P - b \dot{\theta}.$$

Thus the dynamic model of the soft agent is given by

$$V E k^2 \theta + b \dot{\theta} + I \ddot{\theta} = V E c P.$$

This relation models the motion dynamics of the soft agent when a single chamber is actuated. Our parallel soft agent have a total of six identical chambers, therefore similar motion dynamics can be applied to each chamber, although in a rotated reference frame. The multiple chamber actuation can be calculated by superposition of individual chamber actuation.

3.2.3 Actuation and Sensing

The actuation principle of the 6-chambered parallel soft agent is shown in Fig. 3.4. All six chambers inside the soft agent are connected with the air pump through 3-port 3-position solenoid valves. The 3 ports of each valve are connected as: one output port is connected with one of the chambers of the soft agent and the other two ports are connected with the air pump and atmospheric pressure respectively. The 3 positions of the solenoid valves correspond to inward flow, hold the air inside the chamber and outward flow. The solenoid valves cannot be directly driven through microcontroller pins because of the high current requirement. Therefore, the solenoid valves are driven through MOSFET switches, which in turn are controlled through a microcontroller. The expansion of any chamber in the soft agent is proportional to the inward flow duration of the air i.e. the opening time of the solenoid valve.

We used an orientation sensor for measurement of the bending angle. The sensor was mounted on the top of the soft agent. The orientation sensor is used to estimate the amplitude of the oscillation and settling time of the soft agent. Although in this

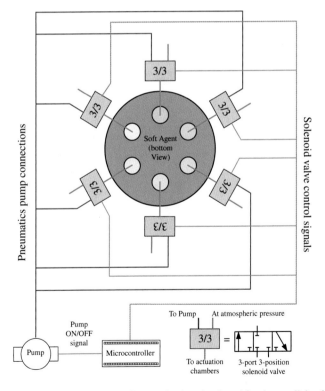

Fig. 3.4 Schematic diagram of the actuation mechanism developed for the parallel soft agent used in our experiments

study we are just concerned about vibration damping of radially opposite chambers, the orientation sensor is capable of measuring 3D rotations of the top surface of the soft agent.

3.2.4 Damping Mechanism

As already explained in Sect. II-A, the previous work in soft agent oscillation damping adds additional mechanical components overhead. In our work, we propose a novel design of the soft actuator, in which the damping effect is achieved by the smart distribution of inflating chambers inside the soft agents. In our 6-chambered parallel soft agent, the number of linear chambers was chosen to be even i.e. six, so that on the circular distribution of chambers, there is always a chamber radially opposite to another chamber i.e. there are always two chambers at 180° from each other as shown in Fig. 3.4. To understand the damping mechanism, refer to Fig. 3.5. For simplicity and ease of explanation, the image just shows 2D planner motion. Suppose, in the current state, the left chamber of the soft agent is actuated and the current pose of the top surface of the soft agent is rightward at an angle of 45°. On deactuation, it will return to vertical position i.e. top surface angle becomes 0. This will require the deflation of the left chamber creating a leftward bending force. If there is no active damping, this force will bend the actuator to the vertical position but with oscillations. Now consider simultaneous actuation of the right chamber, but for a very small period of time. This actuation of the right chamber will produce a smaller rightward bending force. Since the right chamber is actuated only for a smaller duration, the pressure developed inside it will be smaller as compared to deactuation pressure of the left chamber. The net force is still leftward, but the little rightward force is sufficient

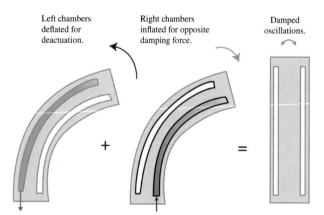

Fig. 3.5 2D illustration of the active oscillation damping mechanism during deactuation. The daming force is created by simultaneous deactuation of left chamber and actuation of right chamber

enough to create oscillation damping effect. The amount of damping is dependent on the ratio of right chamber deactuation pressure and left chamber actuation pressure and is defined as

$$\delta = \frac{\text{damping Actuation Pressure}}{\text{Deactuation Pressure}} \tag{3.4}$$

where δ is damping actuation ratio, and its effect on oscillation damping is analyzed in the Results section.

3.3   Experimental Platform

The experimental platform constructed to perform the experiments is shown in Fig. 3.6. The experimental platform consists of a strong plastic base, on which the soft agent was mounted vertically. All the pneumatics, electrical and electronics systems are attached to the base of the plastic platform, to make the system portable. We used six 3-port 3-position solenoid valves i.e. one for each chamber inside the soft agent. We used an Arduino Uno as the controller board, for sensor data acquisition

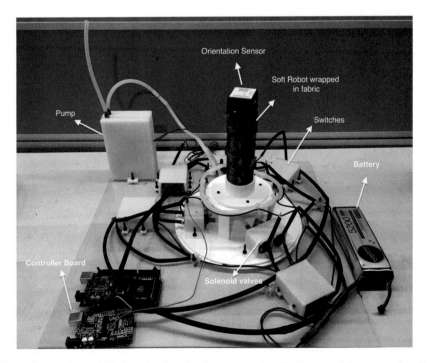

Fig. 3.6 Experimental Platform developed to demonstrate the effectiveness of the proposed oscillation damping method

and processing. To collect vibration information, we used a 3-axis orientation sensor which gives 3D rotations (i.e. roll, pitch, and yaw) of the top surface. The sensor is glued to the top of the soft agent. The orientation sensor was connected to the Arduino Uno using Bluetooth connection. The 3-port 3-position solenoid needs to be driven by high current, which Arduino Uno pins cannot drive directly. To isolate the microcontroller pins from the solenoid valves, we used MOSFET switches. These switches are capable of providing high current required to drive the solenoid valves.

3.4 Experimental Results

In this section, we will report the experimental results to verify the effectiveness of the proposed active oscillation damping method. We will also analyze the effect of damping actuation ratio (δ) defined in (3.4). Figure 3.7 sshows that oscillation profiles under different damping actuation ratio. It can be seen that in the absence of any active damping, the soft agent produce quite large oscillations (about 18° peak value) and settles down to the deactuated position after about 1 second. But when we start to apply the active oscillation damping mechanism as explained in Sect. 3.2.4 the oscillation amplitude began to decay and the settling time becomes small. It can

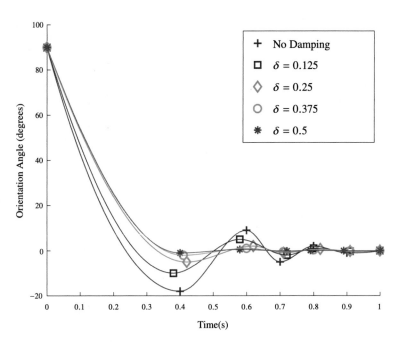

Fig. 3.7 Oscillation profile of the soft agent with no damping compared with the active oscillation damping. The results show that as the damping actuation ratio (δ) increase, the value magnitude of oscillation decay

be seen that the decrease in the oscillation magnitude and settling time is directly proportional to the damping actuation ratio (δ). It is also worth considering that after damping actuation ratio becomes sufficiently large (about 0.5), the system even starts to behave like a nearly overdamped system. The peak oscillation magnitude reduces to a mere $2°$ and settling time reduces to 0.5 seconds. The demonstration of active oscillation damping can be seen at video available on this link: https://www4.comp. polyu.edu.hk/~csahkhan/damping.mp4.

3.5 Conclusion

Due to low stiffness of the soft agents, they exhibit large oscillations on quickly actuations and deactuation. This limits utility in time-critical applications. Previous work for oscillations damping requires additional mechanical components, making system bulky. In this chapter, we presented a novel active damping approach for soft agents, by a smart distribution of actuation chambers inside the soft agent's body. The approach leverage the structure of soft agent itself to actively create damping effect, therefore, requires no additional mechanical and electrical overhead. We proved the efficacy of the proposed method by experimental results.

References

1. H. Lin, F. Guo, F. Wang, and Y.-B. Jia, "Picking up a soft 3d object by "feeling" the grip," *The International Journal of Robotics Research*, vol. 34, no. 11, pp. 1361–1384, 2015.
2. P. Polygerinos, S. Lyne, Z. Wang, L. Nicolini, B. Mosadegh, G. Whitesides, and C. Walsh, "Towards a soft pneumatic glove for hand rehabilitation," in *Intelligent Robots and Systems (IROS), 2013 IEEE/RSJ International Conference on*, pp. 1512–1517, IEEE, 2013.
3. P. Polygerinos, Z. Wang, K. Galloway, R. Wood, and C. Walsh, "Soft robotic glove for combined assistance and at-home rehabilitation," *Robotics and Autonomous Systems*, vol. 73, pp. 135–143, 2015.
4. Y.-L. Park, B.-R. Chen, N. Pérez-Arancibia, D. Young, L. Stirling, R. Wood, E. Goldfield, and R. Nagpal, "Design and control of a bio-inspired soft wearable robotic device for ankle–foot rehabilitation," *Bioinspiration & biomimetics*, vol. 9, no. 1, p. 016007, 2014.
5. I. Galiana, F. Hammond, R. Howe, and M. Popovic, "Wearable soft robotic device for post-stroke shoulder rehabilitation: Identifying misalignments," in *Intelligent Robots and Systems (IROS), 2012 IEEE/RSJ International Conference on*, pp. 317–322, IEEE, 2012.
6. A. H. Khan, S. Li, and X. Luo, "Obstacle avoidance and tracking control of redundant robotic manipulator: An rnn based metaheuristic approach," *IEEE Transactions on Industrial Informatics*, 2019.
7. Y. Zhang, S. Li, J. Zou, and A. H. Khan, "A passivity-based approach for kinematic control of redundant manipulators with constraints," *IEEE Trans. on Ind. Informatics*, 2019.

8. L. Xiao, S. Li, F.-J. Lin, Z. Tan, and A. H. Khan, "Zeroing neural dynamics for control design: comprehensive analysis on stability, robustness, and convergence speed," *IEEE Transactions on Industrial Informatics*, vol. 15, no. 5, pp. 2605–2616, 2018.
9. S. Li, J. He, Y. Li, and M. U. Rafique, "Distributed recurrent neural networks for cooperative control of manipulators: A game-theoretic perspective," *IEEE transactions on neural networks and learning systems*, vol. 28, no. 2, pp. 415–426, 2016.
10. S. Li, S. Chen, and B. Liu, "Accelerating a recurrent neural network to finite-time convergence for solving time-varying sylvester equation by using a sign-bi-power activation function," *Neural processing letters*, vol. 37, no. 2, pp. 189–205, 2013.
11. S. Li and Y. Li, "Nonlinearly activated neural network for solving time-varying complex sylvester equation," *IEEE Transactions on Cybernetics*, vol. 44, no. 8, pp. 1397–1407, 2013.
12. S. Li, Y. Zhang, and L. Jin, "Kinematic control of redundant manipulators using neural networks," *IEEE transactions on neural networks and learning systems*, vol. 28, no. 10, pp. 2243–2254, 2016.
13. S. Li, Z.-H. You, H. Guo, X. Luo, and Z.-Q. Zhao, "Inverse-free extreme learning machine with optimal information updating," *IEEE transactions on cybernetics*, vol. 46, no. 5, pp. 1229–1241, 2015.
14. S. Li, B. Liu, and Y. Li, "Selective positive–negative feedback produces the winner-take-all competition in recurrent neural networks," *IEEE transactions on neural networks and learning systems*, vol. 24, no. 2, pp. 301–309, 2012.
15. J. W. Jeon, "An efficient acceleration for fast motion of industrial robots," in *Industrial Electronics, Control, and Instrumentation, 1995., Proceedings of the 1995 IEEE IECON 21st International Conference on*, vol. 2, pp. 1336–1341, IEEE, 1995.
16. A. H. Khan, Z. Shao, S. Li, Q. Wang, and N. Guan, "Which is the best pid variant for pneumatic soft robots? an experimental study," *IEEE/CAA Journal of Automatica Sinica*, vol. 6, no. 1, p. 1, 2019.
17. L. Jin and S. Li, "Distributed task allocation of multiple robots: A control perspective," *IEEE Transactions on Systems, Man, and Cybernetics: Systems*, vol. 48, no. 5, pp. 693–701, 2016.
18. L. Jin, S. Li, H. M. La, and X. Luo, "Manipulability optimization of redundant manipulators using dynamic neural networks," *IEEE Transactions on Industrial Electronics*, vol. 64, no. 6, pp. 4710–4720, 2017.
19. S. Li, Y. Guo, and B. Bingham, "Multi-robot cooperative control for monitoring and tracking dynamic plumes," in *2014 IEEE International Conference on Robotics and Automation (ICRA)*, pp. 67–73, IEEE, 2014.
20. A. H. Khan, X. Cao, S. Li, V. N. Katsikis, and L. Liao, "Bas-adam: An adam based approach to improve the performance of beetle antennae search optimizer," *IEEE/CAA Journal of Automatica Sinica*, vol. 7, no. 2, pp. 461–471, 2020.
21. A. H. Khan, S. Li, and X. Bin, "Bas-swarm: A nature-inspired metaheuristic algorithm with applications in machine learning," *Soft Computing*, vol. 1, no. 1, p. 1, 2019.
22. E. Pereira, S. S. Aphale, V. Feliu, and S. R. Moheimani, "Integral resonant control for vibration damping and precise tip-positioning of a single-link flexible manipulator," *IEEE ASME Transactions on Mechatronics*, vol. 16, no. 2, p. 232, 2011.
23. C. Laschi, M. Cianchetti, B. Mazzolai, L. Margheri, M. Follador, and P. Dario, "Soft robot arm inspired by the octopus," *Advanced Robotics*, vol. 26, no. 7, pp. 709–727, 2012.
24. A. D. Marchese, R. K. Katzschmann, and D. Rus, "A recipe for soft fluidic elastomer robots," *Soft Robotics*, vol. 2, no. 1, pp. 7–25, 2015.
25. D. Chen, S. Li, F.-J. Lin, and Q. Wu, "New super-twisting zeroing neural-dynamics model for tracking control of parallel robots: A finite-time and robust solution," *IEEE transactions on cybernetics*, 2019.
26. D. Chen, S. Li, W. Li, and Q. Wu, "A multi-level simultaneous minimization scheme applied to jerk-bounded redundant robot manipulators," *IEEE Transactions on Automation Science and Engineering*, 2019.
27. D. Chen, Y. Zhang, and S. Li, "Tracking control of robot manipulators with unknown models: A jacobian-matrix-adaption method," *IEEE Transactions on Industrial Informatics*, vol. 14, no. 7, pp. 3044–3053, 2017.

28. S. Li, Z. Wang, and Y. Li, "Using laplacian eigenmap as heuristic information to solve nonlinear constraints defined on a graph and its application in distributed range-free localization of wireless sensor networks," *Neural processing letters*, vol. 37, no. 3, pp. 411–424, 2013.

29. S. Li, R. Kong, and Y. Guo, "Cooperative distributed source seeking by multiple robots: Algorithms and experiments," *IEEE/ASME Transactions on mechatronics*, vol. 19, no. 6, pp. 1810–1820, 2014.

30. A. T. Khan, S. L. Senior, P. S. Stanimirovic, and Y. Zhang, "Model-free optimization using eagle perching optimizer," *arXiv preprint* arXiv:1807.02754, 2018.

31. Q. Wu, X. Shen, Y. Jin, Z. Chen, S. Li, A. H. Khan, and D. Chen, "Intelligent beetle antennae search for uav sensing and avoidance of obstacles," *Sensors*, vol. 19, no. 8, p. 1758, 2019.

32. A. H. Khan, X. Cao, S. Li, and C. Luo, "Using social behavior of beetles to establish a computational model for operational management," *IEEE Transactions on Computational Social Systems*, vol. 7, no. 2, pp. 492–502, 2020.

33. A. H. Khan and S. Li, "Tracking control of redundant manipulator under active remote center of motion constraints: An rnn-based metaheuristic approach," *SCIENCE CHINA Information Sciences*, 2019.

34. A. H. Khan, S. Li, D. Chen, and L. Liao, "Tracking control of redundant mobile manipulator: An rnn based metaheuristic approach," *Neurocomputing*, 2020.

35. Y. Yang, Y. Chen, Y. Li, M. Z. Chen, and Y. Wei, "Bioinspired robotic fingers based on pneumatic actuator and 3d printing of smart material," *Soft robotics*, vol. 4, no. 2, pp. 147–162, 2017.

36. B. Mosadegh, P. Polygerinos, C. Keplinger, S. Wennstedt, R. F. Shepherd, U. Gupta, J. Shim, K. Bertoldi, C. J. Walsh, and G. M. Whitesides, "Pneumatic networks for soft robotics that actuate rapidly," *Advanced functional materials*, vol. 24, no. 15, pp. 2163–2170, 2014.

37. S. Li and Y. Guo, "Distributed source seeking by cooperative robots: All-to-all and limited communications," in *2012 IEEE International Conference on Robotics and Automation*, pp. 1107–1112, IEEE, 2012.

38. S. Li, Y. Lou, and B. Liu, "Bluetooth aided mobile phone localization: a nonlinear neural circuit approach," *ACM Transactions on Embedded Computing Systems (TECS)*, vol. 13, no. 4, p. 78, 2014.

39. A. H. Khan, S. Li, X. Zhou, Y. Li, M. U. Khan, X. Luo, and H. Wang, "Neural & bio-inspired processing and robot control," *Frontiers in neurorobotics*, vol. 12, 2018.

40. M. Doumit, A. Fahim, and M. Munro, "Analytical modeling and experimental validation of the braided pneumatic muscle," *IEEE transactions on robotics*, vol. 25, no. 6, pp. 1282–1291, 2009.

41. P. Polygerinos, Z. Wang, J. T. Overvelde, K. C. Galloway, R. J. Wood, K. Bertoldi, and C. J. Walsh, "Modeling of soft fiber-reinforced bending actuators," *IEEE Transactions on Robotics*, vol. 31, no. 3, pp. 778–789, 2015.

42. C. W. De Silva, *Vibration damping, control, and design*. CRC Press, 2007.

43. X. Jiang, S. Li, B. Luo, and Q. Meng, "Source exploration for an under-actuated system: A control-theoretic paradigm," *IEEE Transactions on Control Systems Technology*, 2019.

44. Y. Zhang, S. Li, and X. Jiang, "Near-optimal control without solving hjb equations and its applications," *IEEE Transactions on Industrial Electronics*, vol. 65, no. 9, pp. 7173–7184, 2018.

45. X. Jiang and S. Li, "Plume front tracking in unknown environments by estimation and control," *IEEE Transactions on Industrial Informatics*, vol. 15, no. 2, pp. 911–921, 2018.

46. A. T. Khan and S. Li, "A survey on blockchain technology and its potential applications in distributed control and cooperative robots," *arXiv preprint* arXiv:1812.05452, 2018.

47. X. Jiang and S. Li, "Beetle antennae search without parameter tuning (bas-wpt) for multi-objective optimization," *arXiv preprint* arXiv:1711.02395, 2017.

48. F. Ni, A. Henning, K. Tang, and L. Cai, "Soft damper for quick stabilization of soft robotic actuator," in *Real-time Computing and Robotics (RCAR), IEEE International Conference on*, pp. 466–471, IEEE, 2016.

49. Y. Wei, Y. Chen, T. Ren, Q. Chen, C. Yan, Y. Yang, and Y. Li, "A novel, variable stiffness robotic gripper based on integrated soft actuating and particle jamming," *Soft Robotics*, vol. 3, no. 3, pp. 134–143, 2016.

50. Y. Li, Y. Chen, T. Ren, and Y. Hu, "Passive and active particle damping in soft robotic actuators," in *Robotics and Automation (ICRA), 2018 IEEE International Conference on*, pp. 1547–1552, IEEE, 2018.
51. "Smooth-on inc." https://www.smooth-on.com/tb/files/DRAGON_SKIN_SERIES_TB.pdf. Accessed: 2018-08-28.
52. A. Marchese, R. Tedrake, and D. Rus, "Dynamics and trajectory optimization for a soft spatial fluidic elastomer manipulator," *The International Journal of Robotics Research*, vol. 35, no. 8, pp. 1000–1019, 2016.
53. B. Tondu and P. Lopez, "Modeling and control of mckibben artificial muscle robot actuators," *IEEE control systems*, vol. 20, no. 2, pp. 15–38, 2000.

Chapter 4
Management of Electrical Machine Using Torque Control Strategy

Abstract Chapter one of this brief discussed the techniques to control the kinematic motion of the articulated agents in such a way that they avoid collision with obstacles in the surrounding environment. However, chapter one just considers the problem of generating the joint-angles trajectory; it did not discuss how to actually move the joints to the calculated angles. Motors constitute an essential component of articulated agents since they primarily manage the motion of each joint. DC motors are one of the most common types of motors used in articulated agents. For the proper operation of an articulated agent, the lower-level control loops, i.e., the speed-control, current-control, and torque-control loops, must be well-formulated and able to track the reference signal accurately. In this chapter, we present an experimental platform consisting of two motors, mechanically coupled through the shaft, to study the simultaneous management of current and speed in DC motor. We propose the mathematical formulation of the kinematics and dynamics of the system and formulate a Proportional Integral (PI) controller combined with feedforward control law to control the current in the DC motor accurately. The experimental results presented in the chapter show that the bandwidth of the controller depends on the controller parameter and the filtering of the sensor value. If the filtering action is applied to the sensor value, the accuracy is increased, however, it decreases the bandwidth and increases the rise time of the controller. However, by appropriately selecting the filter, a compromise between bandwidth and accuracy can be achieved.

4.1 Introduction

As discussed in Chap. 1, articulated agents plays an important role in several field real-world applications [1–11]. An essential component of articulated agents is motor which is used to manage the motion of its joint. Most of the top-level algorithms, e.g., path planning and obstacle avoidance, assumes that the lower-level control loops are well-tuned and giving an accurate response [12–20]. These lower-level control loops include speed, current, and torque control. In this chapter, we analyze the performance of one of the most commonly used PI controller for regulating

© The Author(s), under exclusive license to Springer Nature Singapore Pte Ltd. 2021   69
A. H. Khan et al., *Management and Intelligent Decision-Making in Complex Systems: An Optimization-Driven Approach*, https://doi.org/10.1007/978-981-15-9392-5_4

the static and dynamic response of the current loop. The function of controlling the current-loop is to regulate the joint-torque of the articulated agent essentially. For the articulated agents used in industry, it is essential to accurately measure and regulate the amount of joint-torque to prevent any damage to the articulated agent [21–31]. Impedance control is a popular control scheme, commonly used, for the control of large articulated agents working in industrial settings [13, 28, 32–41]. A basic principle of impedance control is to regulate the joint-torque instead of joint-velocity or acceleration. This scheme can prevent potential damage to the articulated agent because, in case of velocity control, if an obstacle obstructs the path, the torque will try to get arbitrarily high to maintain the same velocity. Therefore, to regulate the torque, the accurate control of the current loop is essential.

In this chapter, we present an experimental study to analyze the accuracy of the PI controller and the factor affecting it. First, we present the theoretical formulation of the kinematic and dynamic model of the DC motors and then analyze their state-space model. Then we present the design of two experimental platforms; single-motor and dual-motor systems. The purpose of constructing the dual-motor system is to control the current and speed of the motors simultaneously. The first motor is running the current control loop, while the second motor is running the speed control loop. Since the shafts of both motors are coupled, the speed control of the second motor primarily regulates the speed of both motors. In the experimental section, we analyzed the effect of controller parameters and filtering of sensor output on the bandwidth and response rate of the controller.

4.2 DC Motor Model

4.2.1 Dynamic Model

In this section we will derive the dynamic model of a brushed DC motor. Consider a DC motor as shown in Fig. 4.1. The torque produced by the motor is proportional to its current

$$\tau_{motor} = K_t i, \tag{4.1}$$

where K_t is the motor torque constant and i is the rotor current. The other torques applied on the rotor are; frictional torque ($\tau_{friction}$) and sum of torques from external source (τ_{ex}). The frictional torque is propotional to rotor speed and defined as

$$\tau_{friction} = b\dot{\theta},$$

where b is the friction damping constant and θ is the motor rotational angle. According to Newton's 2nd law for rotatioal motion

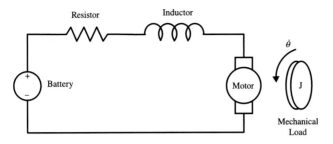

Fig. 4.1 Schematic Model of the DC motor. It shows the connection between the mechanical and electrical parameters of the DC motor

$$J\ddot{\theta} = \sum_k \tau_k$$

$$= \tau_{motor} - \tau_{friction} + \tau_{ex}, \tag{4.2}$$

$$\implies J\ddot{\theta} = K_t i - b\dot{\theta} + \tau_{ex},$$

where J is the rotational inertia of the rotor.

Similarly using Kirchhoffs voltage law, we can formulate the electrical dynamics of the DC motor. According to the Kirchhoff's voltage law the sum of all voltages in a closed loop is zero,

$$\sum_k V_k = 0, \tag{4.3}$$

$$\implies V - Ri - L\frac{di}{dt} - K_e\dot{\theta} = 0,$$

where V is the voltage of DC power supply, R and L are the electrical resistance and inductance of the armature. K_e is the electromotive force constant and the term $K_e\frac{di}{dt}$ represent the back emf according to Lenz's law. For a DC motors, $K_t = K_e$, therefore from now on they will be refered as K. Combining (4.2) and (4.3) the motor model can be sumarized as

$$J\ddot{\theta} - Ki + b\dot{\theta} = \tau_{ex}, \tag{4.4a}$$

$$Ri + L\frac{di}{dt} + K\dot{\theta} = V. \tag{4.4b}$$

Since θ is not a variable of interest for DC motor current control, we define $\omega = \dot{\theta}$. The system (4.4) thus transforms into

$$J\dot{\omega} - Ki + b\omega = \tau_{ex}, \tag{4.5a}$$

$$Ri + L\frac{di}{dt} + K\omega = V. \tag{4.5b}$$

Defining a state vector as $\mathbf{X} = [\omega \ \ i]^T$ and an input vector as $\mathbf{u} = [\tau_{ex} \ \ V]^T$, a state space representation of (4.5) can be written as

$$\dot{\mathbf{X}} = \mathbb{A}\mathbf{X} + \mathbb{B}\mathbf{u},$$
$$\mathbf{y} = \mathbb{C}\mathbf{X},$$

(4.6)

where,

$$A = \begin{bmatrix} -b/J & K/J \\ -K/L & -R/L \end{bmatrix}, \quad B = \begin{bmatrix} 1/J & 0 \\ 0 & 1/L \end{bmatrix} \quad \text{and} \quad C = \begin{bmatrix} 1 & 0 \\ 0 & 1 \end{bmatrix}.$$

To analyse the dynamic response of the DC motor model, lets take the laplace transform of the system (4.6),

$$s\mathbf{X}(s) = \mathbb{A}\mathbf{X}(s) + \mathbb{B}\mathbf{U}(s)$$
$$\mathbf{Y}(s) = \mathbf{X}(s)$$

by eliminating $\mathbf{X}(s)$ from the above system,

$$\mathbf{Y}(s) = \mathbb{C}(s\mathbb{I} - \mathbb{A})^{-1}\mathbb{B}\mathbf{U}(s)$$

where $\mathbb{I}$ is a 2×2 identity matrix and the term $\mathbb{C}(s\mathbb{I} - \mathbb{A})^{-1}\mathbb{B}$ is called the transfer matrix $\mathbb{G}(s)$,

$$\mathbb{G}(s) = \mathbb{C}(s\mathbb{I} - \mathbb{A})^{-1}\mathbb{B}.$$

By putting the values of $\mathbb{A}$, $\mathbb{B}$ and $\mathbb{C}$ in above equation and after simplification we get

$$\mathbb{G}(s) = \frac{1}{(Js + b)(Ls + R) + K^2} \begin{bmatrix} Ls + R & K \\ -K & Js + b \end{bmatrix}.$$

(4.7)

Transfer matrix $\mathbb{G}(s)$ gives 4 transfer funtions relating the two inputs (i.e. τ_{ex} and V) to the two outputs (i.e. ω and i) of the DC motor. The elements of the transfer matrix $\mathbb{G}(s)$ are as follow

$$\mathbb{G} = \begin{bmatrix} \frac{\omega(s)}{\tau_{ex}(s)} & \frac{\omega(s)}{V(s)} \\ \frac{I(s)}{\tau_{ex}(s)} & \frac{I(s)}{V(s)} \end{bmatrix}.$$

4.2.2 Steady State Response

Now lets drive the equation for the steady state response for the the DC motor model in (4.6). In steady state, the system states are constant therefore $\dot{\mathbf{X}} = 0$,

$$0 = \mathbb{A}\mathbf{X}_{ss} + \mathbb{B}\mathbf{u}_{ss},$$
$$\mathbf{y}_{ss} = \mathbb{C}\mathbf{X}_{ss}.$$

Now eliminating $\mathbf{X}_{ss}$ from the above equations we get

$$\mathbf{y}_{ss} = -\mathbb{C}\mathbb{A}^{-1}\mathbb{B}\mathbf{u}_{ss}.$$

Putting values of the matrices $\mathbb{A}$, $\mathbb{B}$ and $\mathbb{C}$,

$$\mathbf{y}_{ss} = \frac{1}{bR + K^2}\begin{bmatrix} R & K \\ -K & b \end{bmatrix}\mathbf{u}_{ss},$$
$$\implies \mathbf{y}_{ss} = \frac{1}{bR + K^2}\begin{bmatrix} R\tau_{ex} + KV \\ -K\tau_{ex} + bV \end{bmatrix}.$$

Writing in term of individual outputs we get,

$$\omega_{ss} = \frac{R\tau_{ex} + KV}{bR + K^2},$$

$$i_{ss} = \frac{-K\tau_{ex} + bV}{bR + K^2}. \tag{4.8}$$

Before analyzing the dynamic control of the DC motor, let us first study the steady state response of the system for a special case.

If $\tau_{ex} = 0$, the (4.8) become

$$\omega_{ss} = \frac{KV}{bR + K^2},$$

$$i_{ss} = \frac{bV}{bR + K^2}.$$

We used a typical small DC motor for our experiments, the estimated value of motor parameters are shown in Table 4.1. Putting the values of all parameters in the above equation we get

$$\omega_{ss} \approx 2.81V,$$
$$i_{ss} \approx 0.015V.$$

The maximum voltage for these motors is 12 Volts. Putting this value in above equation, we get the maximum achieveable value of the outputs are

$$\omega_{ss(max)} \approx 33.72,$$
$$i_{ss(max)} \approx 0.18. \tag{4.9}$$

In this book, our main focus is on the current (torque) control of the DC motor, therefore from now onward we will only consider the current output of the DC motor.

Table 4.1 Parameters for a
DC motor shown used for
experiments

Parameter	Value
R	8.35
K	0.355
b	0.0016
J	0.1

The above equation shows that in the absence of an external torque i.e. $\tau_{ex} = 0$, the maximum steady state current that can flow through the motor's armature is $0.18A$. The small value of current also limits the maximum torque output of the motor according to (4.1). Lets again analyze (4.8) to see is there another way to increase the current output of the DC motor. Lets check the case when $\tau_{ex} < 0$, we define $\tau'_{ex} = -\tau_{ex} > 0$, putting it in (4.8) we get

$$i_{ss} = (K\tau'_{ex} + bV)/(bR + K^2) > bV/(bR + K^2) \qquad (4.10)$$

Here the negative value of the torque means that an external torque is being applied in the direction opposite to the motor rotation. By just applying a negative external torque, it is possible to achieve high current and torque outputs from the DC motor. This fact will later be utilized to improve the performance of current control.

4.3 Current (Torque) Control

In this section we will analyze different methods for the current control in DC motors and discuss the advantages and disadvantages of each method.

4.3.1 Single Motor System

According to the DC motor model in (4.6) the system have a total of two inputs i.e. τ_{ex} and Voltage V. If we consider a single motor system as shown in Fig. 4.2. It only consist of a single DC motor and no external torque is being applied i.e. τ_{ex}. The only input available for the current control is the input voltage V. In this case, according to the (4.9) the maximum steady state current than can be achieved is about 0.18A.

The DC motor experimental platform consists of a small DC motor as shown in Fig. 4.2. The motor is fitted with a magnetic encoder having a resolution of 375 pulses per revolution. L298 dual full-bridge motor driver was used to control the voltage input V to the motor. For current sensing we used ACS712 module having a current sensing range from $-5A$ to $5A$. An arduino mega2560 microcontroller board was used to implement the control algorithms.

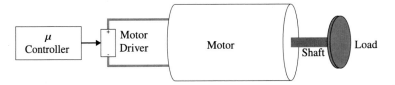

Fig. 4.2 Schematic diagram of the single motor torque control system. The figure shows the mechanical and electrical connection of the motor

4.3.2 Dual Motor System

As discussed in the last section, the single motor system without any external torque cannot achieve very high value of steady state current (torque). High current output requires the presense of an opposing external torque as shown in (4.10). To provide an eternal torque to the motor shaft, we developed a dual motor system as shown in Fig. 4.4. The Primary motor is the one in which we are testing our current control algorithms, whereas the secondary motor is applying an opposing external torque i.e. $\tau_{ex} < 0$. The parameters for dual motor systems were reestimated becuase the attachment of secondary motor will also chnage the dynamic and steady-state characteristics of the primary motor. The estimated parameters are given in Table 4.2. Let us again analyze the motor model (4.5) to see how can we control the secondary motor to provide the required external torque τ_{ex}. The 2nd equation of the motor model states that

$$Ri + L\frac{di}{dt} + K\omega = V.$$

For all practical purposes $L \approx 0$. Therefore

$$Ri + K\omega = V.$$

This equation gives a relation for current i in term of voltage input V and the speed ω,

$$i = \frac{V - K\omega}{R}. \tag{4.11}$$

The relation is plotted in Fig. 4.3 for $V = 12$. The figure show that the as the speed increases the steady state value of the current decrease. Therefore, if we choose a

Table 4.2 Parameters for torque control of dual DC motors system

Parameter	Value
R	8.35
K	0.4485
b	0.002
J	0.1

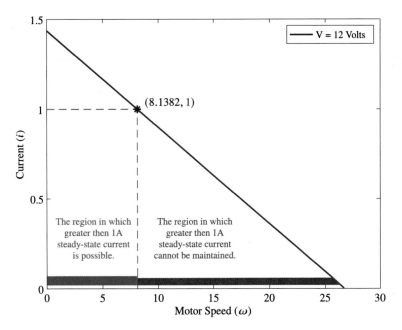

Fig. 4.3 The relation between current and speed of the motor as given (4.11). The graph shows that at very low speed, the motor is capable of providing very high current (torque), however, as the speed increase, the amount of torque provided by the motor decreases

required steady state value of current (say $1A$), it will impose an upper bound on the maximum speed at which the motor can rotate. Consider the following relation

$$i > 1A,$$

$$\implies \frac{V - K\omega}{R} > 1A,$$

by putting the values of parameters and rearranging,

$$\omega < 8.1382 \text{ rad/s.}$$

The above equation shows that if the required value of steady state current is larger then $1A$ then the speed of the motor need to be maintained below 8.1382 rad/sec. Similarly for any given maximum value of a current i_{max}, we have a critical speed ω_c such that

$$\omega < \omega_c.$$

The value of ω_c can is given by

$$\omega_c = \frac{V - i_{ss}R}{K}$$

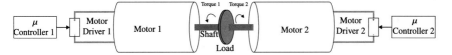

Fig. 4.4 Schematic diagram of the dual motor torque control system. The figure shows the mechanical and electrical connection of the motor. The shaft of both motors is coupled through mechanical load and they apply torque in opposite direction

In our experiments with dual motors, the secondary motor is used to limit the speed of the primary motor below the critical speed ω_c by applying an external torque τ_{ex}, using a PI controller. The control law for the secondary motor is as follow

$$V_{sec} = K_p e(t) + K_i \int_0^t e(t')dt' \tag{4.12}$$

where $e(t) = \omega_c - \omega$. It should be noted that the polarity of the secondary motor should be made such that its direction of rotation is opposite to the primary motor. The block diagram for the dual motor control system is shown in Fig. 4.4.

The experimental platform used for the dual motor system is shown in Fig. 4.4. It consists of 2 small DC motors coupled at shafts. We used two seperate L298N dual full-bridge motor driver, i.e. one for each motor. Although one L298 motor driver module is capable of driving two motors simultaneously but it has a maximum current rating of 2A. Therefore we used seperate L298 motor driver modules for each motor to increase the power capacity of the drivers. Additionaly, using two seperate driver modules provides electrical isolation between motors. In the Fig. 4.4 the motor placed on left is the primary motor. We are testing our current control algorithms on the primary motor therefore a ACS712-5A current sensor is attached with it. The secondary motor is shown on right and running a PI controller (4.12) to always maintain the speed below the critical speed ω_c. A megnatic encoder is attached with the secondary motor to estimate the motors speed. Since shafts of both motors are coupled, therefore we only need one encoder. An Arduino mega2560 controller is used for sensor data acquisition, processing, logging and generation of control signals for both motors.

4.4 Controller Formulation and Performance Analysis

4.4.1 Feed Forward Controller

First we tried a simple PI filter with a constant bias added to it. The bias we calculated using the feedforward kinematic model of the model using

$$V_{bias} = i_{ss}R + K\omega_{ss}, \tag{4.13}$$

where i_{ss} and ω_{ss} are the requried steady state current and speed of the DC motor. We devised the following controller to control the the current of motor 1 in dual motor system

$$V(t) = V_{bias} + K_P e_I(t) + K_I \int_0^t e_I(\tau)d\tau, \tag{4.14}$$

where $V(t)$ controls the input voltage to the motor, $e_I(t) = i(t) - i_{ss}(t)$ is the tracking error of current. Similarly, we applied a PI controller to control the speed of Motor 2. Since the voltage is applied to the motor by controlling the duty cycle of signal from microcontroller, we need to scale and discritize the signal as follow

$$S[n] = \left\lfloor \text{sat}\left(\frac{V(nT_s)}{V_{max}} 2^N \right) \right\rfloor, \tag{4.15}$$

where n is the current time-step, $S[n]$ denotes a digital signal calculated by the microcontroller, V_{max} is the maximum voltage available from motor driver, N is the size of register controlling the duty cycle of PWM, such a register have a total of 2^N possible states. $\lfloor . \rfloor$ function is used to restrict the digial value to integers between $[0, 2^N]$. sat(.) function is defined as follow,

$$\text{sat}(V) = \begin{cases} V, & \text{if } V \leq V_{max} \\ V_{max}, & \text{if } V > V_{max} \end{cases} \tag{4.16}$$

We tested the performance of the controller using the value of: $K_p = 10$ and $K_i = 100$. The steady state reference signal was set to be $i_{ss} = 1A$ and $\omega_{ss} = 5$ rad/s. The first set of experiments was performed by using the raw value from current sensor, i.e., no filtering was applied. Fig. 4.5 shows the profile of current value. It can be seen that the output of the current sensor contains a lot of noise. However, the value of current converges to the reference very quickly. The rise time is around 0.1 seconds which indicate the high bandwidth of the proposed controller. Similarly, the speed profile and value of input duty cycle is also shown in Fig. 4.5. We used

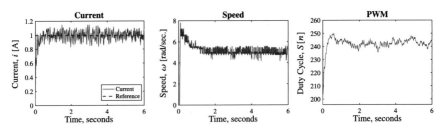

Fig. 4.5 The response of the current control experiment for a constant reference signal using PI controller. The current signal is unfiltered and therefore shows noisy behavior

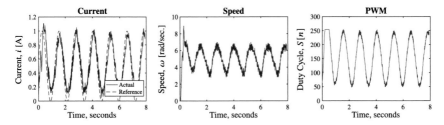

Fig. 4.6 The response of the current control experiment for a constant reference signal using PI controller. The current signal is unfiltered and therefore shows noisy behavior

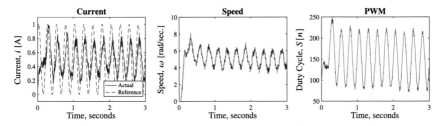

Fig. 4.7 The response of the current control experiment for a constant reference signal using PI controller. The current signal is unfiltered and therefore shows noisy behavior

Arduino Mega 2560 to conduct the experiments, which use an 8-bit ($N = 8$) register to specify the value of duty cycle of PWM. Next we conducted experiments to test the tracking performance of the PI controller. First, we used a sinusoidal reference signal of frequency 5 rad/s. The tracking performance is shown in Fig. 4.6. It can be seen that the controller is easily able to track the input reference signal. The profile of motor speed and pwm signal is also given in the figure. Then we increased the frequency of reference signal to 25 rad/s, and the performance is shown in Fig. 4.7. It can be seen that at this frequency, the amplitude of output starts to decrease and a value of phase difference is also introduced. This shows that the reference frequency is beyond the bandwidth of the controller.

In the previous set of experiments, the raw value of current sensor was used as feedback signal for the PI controller, therefore the output of the current sensor was very noisy. Next we applied a filter to smoothen the sensor output. We used a computationally simple moving avergae filter, which is defined as

$$i_{filter}[n] = \sum_{i=0}^{k} i[n - i], \tag{4.17}$$

where $i[n] = i(nT_s)$ is the discrete value of the current signal and k denotes the order of the filter, i.e., k past value of current sensor are used to calculate the filtered signal. Based on this, the raw value of current in (4.14) is replaced by the filtered value to implement the PI controller. First, we used a constant reference signal of $i_{ss} = 1A$

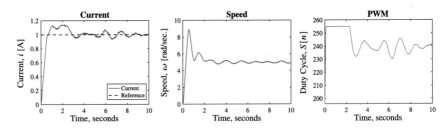

Fig. 4.8 The response of the current control experiment for a constant reference signal using PI controller. The current signal is filtered using the moving average filtering given in, however, it increases the rise time and decrease the tracking bandwidth

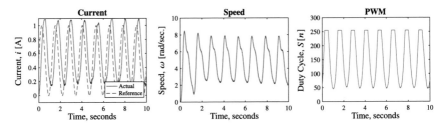

Fig. 4.9 The response of the current control experiment for a constant reference signal using PI controller. The current signal is unfiltered and therefore shows noisy behavior

and $\omega_{ss} = 5$ rad/s. The performance is shown in Fig. 4.8. It can be seen that the filtering of current signal have introduced a delay in the convergence of reference signal and increased the rise time of the output. This indicates that the increase in filtering action have reduced the bandwidth of the PI controller. This is expected, since the moving average is essentially a low-pass filter and therefore it filter out higher frequency, i.e., rapid changes from the output response. Again, we tested the tracking performance and results are shown in Fig. 4.9. The input reference signal was sinusoidal with the frequency of 5 rad/s. It can be seen that decrease in magnitude and phase delay is introduced even at this frequency. It should be noted that for the sinusoidal reference signal with same frequency in case of unfiltered current signal, the controller is able to accurately track as shown in Fig. 4.6.

4.5 Conclusion

In this chapter, we presented the experimental analysis of PI-based control techniques for the torque level control of DC motors for use in articulated agents. The DC motors constitute an essential component for controlling the joints of articulated agents. First, we presented the theoretical model of the DC motors and analyzed its static and dynamic response. Then we gave two types of DC motor systems; single motor and dual motor. The advantage of using a dual-motor system is that we can easily control the speed and current of the DC motor simultaneously. Then we presented experimental results to show the performance of the PI controller using for the dual motor system. Motor 1 was applied a PI controller with the current value as feedback. For motor 2, the speed value was used as feedback of the PI controller. It was shown that for an unfiltered current signal, the rise time is short, and the tracking bandwidth is high. However, the current signal is very noisy. If we apply a moving average filter to the output of the current sensor, the output of current sensor smoothen; however, the bandwidth decreases considerable, and the rise time also increases.

References

1. Y. M. Zhao, Y. Lin, F. Xi, and S. Guo, "Calibration-based iterative learning control for path tracking of ind. robots," *IEEE Trans. on Ind. Electron.*, vol. 62, no. 5, pp. 2921–2929, 2014.
2. C. Yang, Y. Jiang, Z. Li, W. He, and C.-Y. Su, "Neural control of bimanual robots with guaranteed global stability and motion precision," *IEEE Trans. on Ind. Informatics*, vol. 13, no. 3, pp. 1162–1171, 2016.
3. C. Yang, G. Peng, L. Cheng, J. Na, and Z. Li, "Force sensorless admittance control for teleoperation of uncertain robot manipulator using neural networks," *IEEE Trans. on Syst., Man, and Cybern.: Syst.*, 2019.
4. C. Yang, C. Zeng, Y. Cong, N. Wang, and M. Wang, "A learning framework of adaptive manipulative skills from human to robot," *IEEE Trans. on Ind. Informatics*, vol. 15, no. 2, pp. 1153–1161, 2018.
5. H. M. La, T. H. Dinh, N. H. Pham, Q. P. Ha, and A. Q. Pham, "Automated robotic monitoring and inspection of steel structures and bridges," *Robotica*, vol. 37, no. 5, pp. 947–967, 2019.
6. D. Chen, Y. Zhang, and S. Li, "Tracking control of robot manipulators with unknown models: A jacobian-matrix-adaption method," *IEEE Trans. on Ind. Informatics*, vol. 14, no. 7, pp. 3044–3053, 2017.
7. L. Jin, S. Li, X. Luo, Y. Li, and B. Qin, "Neural dynamics for cooperative control of redundant robot manipulators," *IEEE Trans. on Ind. Informatics*, vol. 14, no. 9, pp. 3812–3821, 2018.
8. S. Li, J. He, Y. Li, and M. U. Rafique, "Distributed recurrent neural networks for cooperative control of manipulators: A game-theoretic perspective," *IEEE transactions on neural networks and learning systems*, vol. 28, no. 2, pp. 415–426, 2016.
9. S. Li, S. Chen, and B. Liu, "Accelerating a recurrent neural network to finite-time convergence for solving time-varying sylvester equation by using a sign-bi-power activation function," *Neural processing letters*, vol. 37, no. 2, pp. 189–205, 2013.
10. S. Li and Y. Li, "Nonlinearly activated neural network for solving time-varying complex sylvester equation," *IEEE Transactions on Cybernetics*, vol. 44, no. 8, pp. 1397–1407, 2013.
11. S. Li, Y. Zhang, and L. Jin, "Kinematic control of redundant manipulators using neural networks," *IEEE transactions on neural networks and learning systems*, vol. 28, no. 10, pp. 2243–2254, 2016.
12. Y. Zhang, S. Li, J. Zou, and A. H. Khan, "A passivity-based approach for kinematic control of redundant manipulators with constraints," *IEEE Trans. on Ind. Informatics*, 2019.
13. A. H. Khan, S. Li, and X. Luo, "Obstacle avoidance and tracking control of redundant robotic manipulator: An rnn based metaheuristic approach," *IEEE Transactions on Industrial Informatics*, 2019.
14. L. Xiao, S. Li, F.-J. Lin, Z. Tan, and A. H. Khan, "Zeroing neural dynamics for control design: comprehensive analysis on stability, robustness, and convergence speed," *IEEE Transactions on Industrial Informatics*, vol. 15, no. 5, pp. 2605–2616, 2018.

15. S. Li, S. Chen, B. Liu, Y. Li, and Y. Liang, "Decentralized kinematic control of a class of collaborative redundant manipulators via recurrent neural networks," *Neurocomputing*, vol. 91, pp. 1–10, 2012.
16. L. Jin, S. Li, H. M. La, and X. Luo, "Manipulability optimization of redundant manipulators using dynamic neural networks," *IEEE Trans. on Ind. Electron.*, vol. 64, no. 6, pp. 4710–4720, 2017.
17. S. Li, Z.-H. You, H. Guo, X. Luo, and Z.-Q. Zhao, "Inverse-free extreme learning machine with optimal information updating," *IEEE transactions on cybernetics*, vol. 46, no. 5, pp. 1229–1241, 2015.
18. S. Li, B. Liu, and Y. Li, "Selective positive–negative feedback produces the winner-take-all competition in recurrent neural networks," *IEEE transactions on neural networks and learning systems*, vol. 24, no. 2, pp. 301–309, 2012.
19. L. Jin and S. Li, "Distributed task allocation of multiple robots: A control perspective," *IEEE Transactions on Systems, Man, and Cybernetics: Systems*, vol. 48, no. 5, pp. 693–701, 2016.
20. L. Jin, S. Li, H. M. La, and X. Luo, "Manipulability optimization of redundant manipulators using dynamic neural networks," *IEEE Transactions on Industrial Electronics*, vol. 64, no. 6, pp. 4710–4720, 2017.
21. S. Ling, H. Wang, and P. X. Liu, "Adaptive fuzzy dynamic surface control of flexible-joint robot systems with input saturation," *IEEE/CAA Journal of Automatica Sinica*, vol. 6, no. 1, pp. 97–107, 2019.
22. H. Wang, P. X. Liu, X. Zhao, and X. Liu, "Adaptive fuzzy finite-time control of nonlinear systems with actuator faults," *IEEE transactions on cybernetics*, 2019.
23. H. Wang, P. X. Liu, X. Xie, X. Liu, T. Hayat, and F. E. Alsaadi, "Adaptive fuzzy asymptotical tracking control of nonlinear systems with unmodeled dynamics and quantized actuator," *Information Sciences*, 2018.
24. C. Yang, G. Peng, Y. Li, R. Cui, L. Cheng, and Z. Li, "Neural networks enhanced adaptive admittance control of optimized robot–environment interaction," *IEEE transactions on cybernetics*, vol. 49, no. 7, pp. 2568–2579, 2018.
25. S. Li, Y. Guo, and B. Bingham, "Multi-robot cooperative control for monitoring and tracking dynamic plumes," in *2014 IEEE International Conference on Robotics and Automation (ICRA)*, pp. 67–73, IEEE, 2014.
26. S. Li and Y. Guo, "Distributed source seeking by cooperative robots: All-to-all and limited communications," in *2012 IEEE International Conference on Robotics and Automation*, pp. 1107–1112, IEEE, 2012.
27. S. Li, Y. Lou, and B. Liu, "Bluetooth aided mobile phone localization: a nonlinear neural circuit approach," *ACM Transactions on Embedded Computing Systems (TECS)*, vol. 13, no. 4, p. 78, 2014.
28. A. H. Khan, S. Li, X. Zhou, Y. Li, M. U. Khan, X. Luo, and H. Wang, "Neural & bio-inspired processing and robot control," *Frontiers in neurorobotics*, vol. 12, 2018.
29. X. Jiang, S. Li, B. Luo, and Q. Meng, "Source exploration for an under-actuated system: A control-theoretic paradigm," *IEEE Transactions on Control Systems Technology*, 2019.
30. Y. Zhang, S. Li, and X. Jiang, "Near-optimal control without solving hjb equations and its applications," *IEEE Transactions on Industrial Electronics*, vol. 65, no. 9, pp. 7173–7184, 2018.
31. X. Jiang and S. Li, "Plume front tracking in unknown environments by estimation and control," *IEEE Transactions on Industrial Informatics*, vol. 15, no. 2, pp. 911–921, 2018.
32. H. C. Liaw and B. Shirinzadeh, "Robust generalised impedance control of piezo-actuated flexure-based four-bar mechanisms for micro/nano manipulation," *Sensors and Actuators A: Physical*, vol. 148, no. 2, pp. 443–453, 2008.
33. Q. Xu, "Adaptive discrete-time sliding mode impedance control of a piezoelectric microgripper," *IEEE Transactions on Robotics*, vol. 29, no. 3, pp. 663–673, 2013.
34. W. He, Y. Dong, and C. Sun, "Adaptive neural impedance control of a robotic manipulator with input saturation," *IEEE Transactions on Systems, Man, and Cybernetics: Systems*, vol. 46, no. 3, pp. 334–344, 2016.

35. Q. Xu, "Robust impedance control of a compliant microgripper for high-speed position/force regulation.," *IEEE Trans. Industrial Electronics*, vol. 62, no. 2, pp. 1201–1209, 2015.

36. A. T. Khan and S. Li, "A survey on blockchain technology and its potential applications in distributed control and cooperative robots," *arXiv preprint* arXiv:1812.05452, 2018.

37. X. Jiang and S. Li, "Beetle antennae search without parameter tuning (bas-wpt) for multi-objective optimization," *arXiv preprint* arXiv:1711.02395, 2017.

38. D. Chen, S. Li, F.-J. Lin, and Q. Wu, "New super-twisting zeroing neural-dynamics model for tracking control of parallel robots: A finite-time and robust solution," *IEEE transactions on cybernetics*, 2019.

39. D. Chen, S. Li, W. Li, and Q. Wu, "A multi-level simultaneous minimization scheme applied to jerk-bounded redundant robot manipulators," *IEEE Transactions on Automation Science and Engineering*, 2019.

40. Q. Wu, X. Shen, Y. Jin, Z. Chen, S. Li, A. H. Khan, and D. Chen, "Intelligent beetle antennae search for uav sensing and avoidance of obstacles," *Sensors*, vol. 19, no. 8, p. 1758, 2019.

41. A. H. Khan, Z. Shao, S. Li, Q. Wang, and N. Guan, "Which is the best pid variant for pneumatic soft robots? an experimental study," *IEEE/CAA Journal of Automatica Sinica*, vol. 6, no. 1, p. 1, 2019.

Printed in the United States
By Bookmasters